Global Research Ethics

Global Research Ethics is a guide for students and their instructors as well as practitioners and researchers to understand topics linked to research ethics from a more global perspective. Research plays a key role in identifying health disparity trends and evaluating interventions to improve the health and well-being of the populations at the individual, local, national, and global levels. Conducting ethically sound research is imperative in these contexts.

This book (a) uses case studies to offer examples of current research ethical dilemmas and (b) considers regulatory and cultural frameworks in a number of country contexts that highlight diverse methods of identifying and managing these ethical dilemmas. Chapters cover different types (groups) of participants, issues in research, and ways of doing research; then each chapter looks at exemplar case studies with at least two analytical commentaries. Case studies include health and social care research, and originate from countries such as Brazil, Chile, South Africa, Botswana, Australia, and New Zealand, as well as the U.S. and U.K. The different viewpoints showcased will allow for dialogue to ensue about the ways in which populations and topics in research need to be conceptualized.

Global Research Ethics is suitable for all undergraduates and postgraduates in research methods courses in the social and health sciences. It provides academic researchers, students, and community partners with guidelines to reflect on as they develop their own research studies.

Caren J. Frost is an Associate Vice President for Research Integrity & Compliance (AVPRIC) and a Research Professor in the College of Social Work at the University of Utah, U.S.A. Her areas of expertise are in qualitative research methods, women's health, and research integrity.

Lisa H. Gren is a mixed methods researcher working in Utah, U.S.A., and globally. She focuses on collaborative approaches to systems changes that are responsive to health concerns identified by community members. She served as an IRB reviewer for eight years and trains graduate students on ethical research practices in community settings.

L. Scott Benson is an infectious disease trained physician who also holds a doctoral degree in Civil and Environmental Engineering. He has been working globally for over 20 years in the Dominican Republic, Peru, Ghana, and Pakistan. His main focus is the spread of infectious diseases to human populations from the environment.

Margaret Carlson is currently working on her doctorate in Public Health and has worked in clinical research for the past seven years. Her experience is in both adult and pediatric research with a background in regulatory administration and implementation.

Global Research Ethics

Case Studies from International Research Contexts

Caren J. Frost, Lisa H. Gren, L. Scott Benson, and Margaret Carlson

Routledge
Taylor & Francis Group
LONDON AND NEW YORK

Designed cover image: Andriy Onufriyenko / Getty Images

First published 2023
by Routledge
4 Park Square, Milton Park, Abingdon, Oxon OX14 4RN

and by Routledge
605 Third Avenue, New York, NY 10158

Routledge is an imprint of the Taylor & Francis Group, an informa business

British Library Cataloguing-in-Publication Data
A catalogue record for this book is available from the British Library

ISBN: 978-0-367-47078-4 (hbk)
ISBN: 978-0-367-47079-1 (pbk)
ISBN: 978-1-003-03323-3 (ebk)

DOI: 10.4324/9781003033233

Typeset in Bembo
by Newgen Publishing UK

Access the Instructor and Student Resources: https://www.routledge.com/9780367470791

Contents

Contributors

Qosay Al-Balas, PhD
Jordan University of Science & Technology
Jordan

Roberto Andorno, PhD
University of Zurich, Institute of Biomedical Ethics & Medical History
Switzerland

Anish K.R., PhD
Rajagiri College of Social Sciences
India

Daniel Ansong, MD
Komfo Ankoye Teaching Hospital
Ghana

Berna Arda, MD, PhD
Ankara University
Turkey

Muhammet Arslan, PhD
Hacettepe University, Department of Medical History and Medical Ethics
Turkey

Eunice Bajomo
University of Utah
Nigeria

Gehanath Baral, MD
Kathmandu University, Nobel Medical College
Nepal

Jahn Barlow, MPA
University of Utah
United States

L. Scott Benson, MD, PhD, MPH
University of Utah
United States

Alahí Bianchini, MA
Bioethics Program of FLACSO
Argentina

Edna Bosire, PhD
University of the Witwatersrand (South Africa)
Kenya

Jeffrey R. Botkin, MD, MPH
University of Utah
United States

Caroline Brall, PhD
Eidgenössische Technische Hochschule Zurich
Switzerland

Caroline Brooks
Brighton and Sussex Medical School
United Kingdom

Nüket Örnek Büken, PhD
Hacettepe University, Department of Medical History and Medical Ethics
Turkey

Margaret Carlson, MPH
University of Utah
United States

Yali Cong, PhD
Peking University
China

Lydia Debrah
Ensign College of Public Health
Ghana

Tsivya Jeannette Devereaux
University of Utah
Paraguay

Harneet Kaur Dhillon
University of Utah
India

Sinegugu Evidence Duma, PhD
University of KwaZulu-Natal
South Africa

Agata Ferretti, MA, MSc
Eidgenössische Technische Hochschule Zurich
Switzerland

Caren J. Frost, PhD, MPH
University of Utah
United States

Sadie Gabler
University of Utah
United States

Nishakanthi Gopalan, PhD, MSc
University of Malaysia
Malaysia

Joanna Gorka, MSc, PhD
University of Utah
Croatia/Poland

Doris Gray, PhD
Al Akhawayn University
Morocco

Lisa H. Gren, PhD, MSPH
University of Utah
United States

Felicitas Holzer, PhD
University of Zurich
Switzerland

Charles P. Hoy-Ellis, PhD, MSW
University of Washington
United States

Ronak Iqbal, MD MPH
Practitioner
Pakistan

Valerie Jenkins, DPhil
University of Sussex
United Kingdom

Ng Chirk Jenn, MBBS, MMed, PhD
University of Malaysia
Malaysia

Hena John Fisk, PhD
Practitioner
India

Erin P. Johnson, PhD
University of Utah
United States

Jordan Johnson, MPH
University of Utah
United States

Sarmishta Diraviam Kannan
University of Utah
India

Tomas Klingström, PhD
Swedish University of Agricultural Sciences
Sweden

Jorge E. Lemus, PhD
Universidad Don Bosco
El Salvador

Fernando Lolas, MD
University of Chile
Chile

Weihua Lui, MPH
Wuhan University of Technology
China

Zamasomi Prudence Busisi Luvuno, PhD
University of KwaZulu-Natal
South Africa

Aileen Savina Marques, BA, LL MS
Advocate
India

Diana Mendieta
Catholic University of Asuncion
Paraguay

Rebeca A. Michael, MPH
University of Utah
United States

Yohannie Mlombe, MD
Malawi College of Medicine
Malawi

Busisiwe Purity Ncama, PhD
University of KwaZulu-Natal
South Africa

Katelyn O'Farrell, MS
University of Utah
United States

Simin Peng, MSW
Shenzhen East West Social Work Service Agency
China

Ginu George Powathil, PhD
Rajagiri College of Social Sciences
India

Bridget Pratt, MD
University of Melbourne
Australia

Courtney Rada
University of Utah
Hungary

Brenda Ralls, PhD
Utah Department of Health
United States

Eduardo Rodríguez, PhD
University of Chile
Chile

Erin Rothwell, PhD, MS
University of Utah
United States

Luisa Rusta
University of Utah
Germany

Mohammed Sbai, PhD, EMS
Moulay Ali Institute for Rehabilitation
Morocco

Preethi Shivayogi, PhD
Lifeline Ethics Committee
India

Hannah Simonds, MBChB, MRCP
Tygerberg Academic Hospital
South Africa

O. Fahina Tavake-Pasi, MS
Executive Director at National Tongan American Society
Tonga

Ida Tovar
University of Utah
United States

Patrick L. Umeh, MPH, BIT
Ensign College of Public Health (Ghana)
Nigeria

Peter West-Oram, PhD, MA
University of Sussex
United Kingdom

Abdullah Yildiz, MD, PhD
Ankara University
Turkey

Bobby J. Younce, MSW, LCSW
University of Utah
United States

Oumaima Zedgui, MS
Moulay Ali Institute for Rehabilitation
Morocco

Huiqin Zhong
Shenzhen East West Social Work Service Agency
China

Preface

Research plays a key role in identifying health disparities and trends and evaluating interventions to improve the health and well-being of the populations at the individual, local, national, and global levels. Conducting ethically sound research is an international imperative. Although the basic principles when conducting ethical research are shared among countries, there is room for improvement to create open conversations among research communities to improve research practices. Further exploration and collaboration are needed to (a) develop a cohesive view on ethical research and (b) advance conversations and guidelines created to address emerging issues (Bhutta, 2002).

This book aims to be a guide for individuals to understand research and ethics from a more global perspective. From discussions with professionals from all over the world, there is a need for further sharing of information about how to manage cross-cultural ethics issues in research contexts. Community and academic partners at local levels are sometimes not able to conceptualize how to conduct rigorous, non-coercive, and confidential research. This book will use case studies and offer examples of ethical issues and solutions in research to address some situations. It will provide academic researchers, students, and community partners with clear parameters and guidelines to ensure they are conducting research in an ethical manner that fits with global considerations of these issues and structures. There are numerous ways community groups understand ethical frameworks for conducting research. These different views should be a point of dialogue among partners so that the complexity of ethics can be managed together on a research project. Discussants from six continents and various countries will provide the information on global research ethics for this book.

Ethical issues are found continually in the news and this information aids the need for this book and what it hopes to do. For example, human genome research has made headlines over the years and recently it was brought up again outlining the ethical misuse of data sample gathering from Indigenous communities by Western scientists and the mistrust that created. One scientist was trying to change this relationship by a funded program called Summer Internship for Indigenous Peoples in Genomics. Funded by the U.S. National Institutes of Health (NIH), this program hoped to train individuals in genomics so they can "introduce that field's tools to their communities as well as bring the Indigenous perspective to research" (Wade, 2018). Another topic is euthanasia and the legalization of assisted suicide and the populations it can encompass. In Canada, physicians and other ethicists have proposed the use of euthanasia as an option for children, as well as an option for those suffering from a mental illness (Van Maren, 2018). Another controversial topic of the use of medical cannabis has been spreading throughout Europe with more countries legalizing its use. However, the lack of information and research behind the use of medical cannabis, along with the ethics of using it have created discussions throughout Europe on this topic (Murphy, 2018).

This book links to courses on global research and global ethics for upper-level undergraduate, lower-level graduate students, and community partners. Having access to information about how research is managed in other countries and with individuals and/or communities from those countries will enable researchers in various job roles and collaborating with various communities to have a more comprehensive understanding of how to manage the ethical situations and issues that arise in health sciences, medical, and social sciences research activities.

Each chapter will highlight a specific population that has been the center of several ethical issues and debates. We will begin the book with chapters separately dealing with youth, women, and men, and then the following chapters will target aspects of research about language and participation. The last four chapters call attention to high-profile themes of research: economics and incentives, genetics and bio-specimens, mental health, and use of technology. Each chapter will follow a similar outline containing background on the topic, case studies specific to the issue, and responses from professionals from different communities or countries on the case study and issue at hand. In addition, exercises to use in a classroom setting are provided for each chapter. These exercises could be used as conversation starters to highlight the issues under consideration. An appendix of videos and a list of references at the end of the book can be used to summarize and supplement the information for each chapter.

1 Introduction to Global Research Ethics

Introduction

Over the past 20 years, global, interdisciplinary research has increased and is becoming more prevalent. Many universities have connections with academic settings in other countries and on other continents. Consequently, research ethics parameters need to be reconsidered and reframed based on different cultural understandings of how to interact with human participants in research settings and how to develop cross-national research studies. This book will provide insights into topics linked to global research ethics for work with specific types of populations and on specific research areas that need a more global discussion. This chapter provides the frame for this book, as well as information about the book's structure, which uses case studies and case analyses. This text will be particularly useful in highlighting considerations for global, interdisciplinary work so that research ethics will become a regular and ongoing component for colleagues working transnationally.

The U.S. Department of Health and Human Services defines research as "a systematic investigation, including research development, testing and evaluation, designed to develop or contribute to generalizable knowledge" (USDHHS, 2019). Human subject is defined as "a living individual about whom an investigator (whether professional or student) conducting research obtains data through intervention or interaction with the individual, or identifiable private information" (USDHHS, 2019). The National Institutes of Health define ethics as "the norms for conduct that distinguish between acceptable and unacceptable behavior" (Resnik, 2015; BBC, 2014). Ethics help govern research in multiple ways by setting guidelines and rules holding those conducting research responsible for their actions by ensuring their work does not violate the rights of participants (i.e., human subjects). With globalization, research cannot stay within a nation's borders and is best addressed by sharing data, skills, and education with others. According to the Merriam-Webster dictionary, global is defined as relating to or involving the entire world (Merriam-Webster, 2017). Research and ethics reflect this global nature in that nations over the world have come together to produce documents and guidelines to help aid and standardize research going forward. Several of these documents will be mentioned later in this chapter since they helped outline practices over the years and how research is shaped today.

DOI: 10.4324/9781003033233-1

Human Rights and Research Ethics: The Foundational Documents

The treatment of humans in World War II sheds light on the inhuman medical and research practices that prisoners and other populations were subjected to and called attention to the need to protect human rights. It is well documented that during this time human beings were subjected to experimentation without being allowed to consent to such experimentation and without regard to risks to their physical and emotional well-being. Experimentation was justified with the aims of bettering survival for military personnel, expanding and perfecting the health of some groups over others, and/or advancing pharmaceutical options and specific treatments (USHMM, 2019a). The world was shocked that individuals in power in a modern society could allow and participate in such unethical mistreatment with little or no regard for the rights of human subjects (O'Mathuna, 2006).

Following World War II, 13 trials were conducted between 1945 and 1949 by the Allied Forces to prosecute members and leaders of Nazi Germany who participated in and allowed for mistreatment to occur. Known as the Nuremberg Trials, after the city in which they were held, 185 people were indicted and 23 of these individuals were charged with crimes against humanity, such as medical experimentation (History Channel, 2010). The trials were broken into two parts: the major war criminals' trial and subsequent trials. The major war criminals' trials consisted of indictments of members and leaders of the Nazi Party and the Gestapo (History Channel, 2010). The subsequent trials included charges of crimes against humanity and included trials of German doctors, judges, industrialists, and military officers. During the Doctors' Trial, it was argued these medical professionals did nothing outside of the scope of work that was not already being done elsewhere in the world. The Trials also pointed to the fact that there was no official documentation or law in countries and/or at the global level that stated what could be done during research activities and experimentation using human participants (USHMM, 2019b). These allegations led the way for ethical codes for research and medical practice to be developed.

The Nuremberg Code was developed in 1947 based on information from these trials (Fischer, 2006). This Code was the first set of standards that outlined the rights humans have when participating in research to prevent future abuse, and included concepts of morality, ethics, and legal precautions (USHMM, 2019a). Comprising ten points, a main emphasis of the Nuremberg Code is that research must be voluntary for an individual to participate. An individual could not be forced or coerced into participating, and people should not feel under pressure to participate. Other points more specifically outlined what research was to entail: how it must be useful and must have more benefits than risks to the participant and society (Fischer, 2006). Thus, research and experiments with human participants should be conducted for the good of society, and not for personal study or amusement. In addition, personnel conducting such studies should be scientifically qualified in their scientific discipline of expertise. The Nuremberg Code provided the basis for the further development of standards and protocols for conducting research with human participants, and it has been used as a basis for agreements when participating in research today, especially around informed consent parameters.

After the founding of the United Nations (UN) in 1945, a second foundational document was developed by the UN Commission on Human Rights (UN, 2018). This Commission, headed by Eleanor Roosevelt, created a declaration of rights that was completed in 1948 and was based on the findings from the Nuremberg Trials. Forty-eight of the 58 UN member countries voted in favor of this document depicting the rights all people should have—"the first time countries agreed on a comprehensive statement of inalienable human rights" (UN, 2018). This document, titled the Universal Declaration of Human Rights (UDHR), was created to be a universal set of standards for defining what it meant to be a human being for all people, regardless of gender, race, location, etc. and "remains the primary source of global human rights standards" (Hannum, 1998). Despite the small number of historical voting members, the UDHR has been translated to over 500 languages (see Figure 1.1). Currently, with 192 member states, the UDHR is widely upheld and used, because it was a landmark achievement in defining and standardizing the rights of all humans, is part of the International Bill of Human Rights (UN, 2018).

The UDHR outlines 30 rights that are grouped into seven categories, some of which link to conducting research. These categories cover civil, political, life, liberty, property, speech, privacy, economic, social, and cultural rights (UN, 2018). In addition, several of the UDHR articles reference research specifically. For example, Article 5 states, "no one shall be subjected to torture or to cruel, inhuman, or degrading treatment" (UN, 2018). Articles 18 and 19 highlight rights of freedom of thought and opinion or expression, which are key elements of an informed consent process and discussion in invitations to participate in research studies. Finally, Article

Figure 1.1 Covers of the Universal Declaration of Human Rights

25 states the right to medical care and necessary social services, which should also be represented in an informed consent process.

Although it is not legally binding, the UDHR has been used in the construction of laws, treaties, and constitutions (UN, 2018). The UN uses the UDHR as a basis for wording of the UN Charter, which is, in fact, binding for all 193 member states. The UDHR has been used to build cases against parties that break the rights outlined within it and as an instrument in helping countries define and implement their own human rights agendas (UN, 2018). While not considered international law, the UDHR has been used and upheld in international relations. It is because of its general references and open definitions that it can be widely used and accepted as a standard declaration of rights and is pertinent for research structures (Hannum, 1998).

Several other documents were developed after World War II and should be considered for understanding global research ethics. Assembled in 1945, the World Medical Association (WMA) was concerned by the "treatments" physicians gave to patients and created a code of medical ethics known as the Declaration of Geneva in 1949, which outlined principles a physician should uphold when working with patients (WMA, 2018; Fischer, 2006). Elaborating on this Declaration, the WMA wanted to provide more specific information and interpretation on the responsibilities for a physician when conducting research. Thus, the *Ethical Principles for Medical Research Involving Human Subjects*, later known as the Declaration of Helsinki (DoH) in 1964, was published (Fischer, 2006). The DoH's purpose, like that of the Nuremberg Code, was to prevent any mistreatment of human subjects in research. This document puts the responsibility of the welfare of research participants on the medical provider and researcher; however, it allows for research participant responsibility as well. The three main sections of the DoH are (WMA, 2018):

1. Description of why research involving human participants is needed—this section discusses vulnerable populations, the need to provide special care for those populations when conducting research with them, and the requirement of obtaining consent from all participants;
2. Reiteration of the Nuremberg Code—this section expands on the construct of voluntary participation; stipulates the need to inform participants about the risks, possible benefits, and ability to stop study participation; and outlines the importance of only conducting research with participants who might benefit from it, not just arbitrary individuals to obtain sufficient sample size; and
3. Discussion about how research and medical care are connected and that participants may not understand these connections—this section addresses the use of a placebo to test out possible new methods of treatment and examines the concept of explaining the use of placebo to participants (USDHHS, 2017).

The most important aspect of the DoH is that it indicates the importance of obtaining informed consent for individuals participating in research studies (Fischer, 2006). Informed consent not only applies to adults, but to vulnerable populations such as children, prisoners, people with mental challenges, and military personnel

(Fischer, 2006). Several adaptations and revisions have been made to the DoH over the years but the emphasis on informed consent and the well-being of participants, rather than conducting experimentation on humans in the sole interest of science, has stayed consistent. This emphasis on the public health of a community, as opposed to an individual and their welfare, is a different condition compared to previous documents (Williams, 2008). This shift not only affected how research was approached and how individuals were enrolled, but also affected how research findings could be published. Part of the new revisions made to the DoH outlined policies for publication of research that require a review of the research to be conducted by a committee to ensure the rights of the research participants were upheld and put before the needs of society (Williams, 2008).

Amendments made to the DoH prior to 2000 kept this emphasis of the individual over the health of the public (Williams, 2008). In 2000, the updated version of the DoH outlined an emphasis on the benefits of future patients instead of only current participants. This trend toward including more public health-based objectives continued in the 2008 amendment. In this version, further inclusion of public benefit is made by introducing epidemiologic research (the "study of populations to understand the causes and patterns of health and illness") into the discussion (USDVA, 2016). This version also included wording to include communities in addition to individuals when discussing the risks of a research study. Although this amendment is moving toward more of a public health and community-based approach, the rights and welfare of the individual are still first and foremost. Thus, the discussions about participants and the purposes of research are evolving, and this book will address some of these critical changes and explore potential challenges for global research ethics in the twenty-first century.

In the late twentieth century, it became apparent that these discussions about how to ensure that research was done ethically and with as little harm to participants as possible, are not just local or national concepts—they have become global discussions. In 1982, the Council for International Organizations of Medical Sciences (CIOMS) along with the World Health Organization (WHO) developed guidelines to help judge the ethics of research. This guideline collaboration was created to add to the foundational documents to be used for more cross-national and cross-cultural research studies (Kim, 2012). Institutional Review Boards (IRBs) or Research Ethics Committees (RECs) help ensure these guidelines are followed. IRBs/RECs are committees composed of scientists and community members who review research studies to ensure that the research parameters are ethical and that the rights of participants are protected. These institutions are the designated bodies of research ethics in over 90 countries. They not only make sure the research is ethical, but they also check for potential biases and conflicts of interest, and ensure research studies adhere to the regulations put into place at local and national levels (Grady, 2015). Personnel involved with the study are upheld to guidelines, and in some cases may require certification by a governing body, to conduct the research. One of these guidelines is good clinical practice (GCP). GCP is an international standard for all aspects of research studies to ensure that procedures are conducted in accordance with ethical standards and current knowledge. The International Conference on

Harmonization-Good Clinical Practice (ICH-GCP) guidelines were created to limit the possible inconsistencies that could occur when research is conducted in different countries. The entities involved in the development of the ICH-GCP were the European Union, Japan, United States, Australia, Canada, Nordic countries, and WHO (Vijayananthan & Nawawi, 2008). There are 13 principles of GCP—all of which aim to hold clinical trials to ethical principles and scientific integrity.

Finally, in 1992 the International Commission on Occupational Health (ICOH) created the International Code of Ethics (revised in 2014) to outline responsibilities that medical practitioners have toward their fellow workers and to the public. The official languages of this Code are English and French; however, the Code is internationally recognized and has been translated into eight additional languages (ICOH, 2019). The ICOH specifically pertains to responsibilities that health professionals have and promotes the need to conduct research for educational and scientific purposes (see table 1.1).

Also of interest is the link below that will take you to a compiled list of documents that pertain to specific countries as well as documents that are globally followed and adhered to: https://www.hhs.gov/ohrp/sites/default/files/2018-International-Compilation-of-Human-Research-Standards.pdf.

Table 1.1 Pertinent global ethics documents

Document Name	*Year Created (Current Revision Year)*	*Weblink*
Nuremberg Code	1947	https://history.nih.gov/research/downloads/nuremberg.pdf
Universal Declaration of Human Rights	1948	https://www.ohchr.org/EN/UDHR/Documents/UDHR_Translations/eng.pdf
Declaration of Geneva	1949 (revised last 2006)	https://www.wma.net/what-we-do/medical-ethics/declaration-of-geneva/
Declaration of Helsinki	1964 (revised last 2008)	https://www.wma.net/policies-post/wma-declaration-of-helsinki-ethical-principles-for-medical-research-involving-human-subjects/
Belmont Report	1979	https://www.hhs.gov/ohrp/regulations-and-policy/belmont-report/read-the-belmont-report/index.html
Good Clinical Practice	1996	https://www.ich.org/fileadmin/Public_Web_Site/ICH_Products/Guidelines/Efficacy/E6/E6_R1_Guideline.pdf
International Code of Ethics	1992 (revised in 2014)	http://www.icohweb.org/site/multimedia/code_of_ethics/code-of-ethics-en.pdf

2 Minors

Research that includes children (minors between the ages of birth and 17 years of age) requires special attention to obtaining assent, parental permission, and/or consent. Vulnerability, lack of understanding about the consequences and implications of research study participation, and the lack of evidence about how to include children as participants are key elements for researchers to consider. Youth are not miniature adults, making research in this field imperative for improving their health outcomes. They may not be able to fully grasp the situation in decision making or deciding to participate in research and may be easily coerced; thus, special precautions are required for their participation. In 1989, the Office of the United Nations High Commissioner for Human Rights (UNCHR) created and ratified the Convention on the Rights of the Child. This Convention is concerned with the rights minors should have in any given situation (UNHCR, 2018). Approximately, 200 entities have signed and/or ratified this Convention. It has been translated into six major languages and comprises 54 articles with several applying to ethical considerations for minors in research (UNCHR, 1995). Article 1 defines the ages for which an individual is considered a child—individuals under the age of 18. Article 3 expresses the extra care and protection that is needed when dealing with children, including information on standards of safety and health. Articles 12, 13, and 14 focus on the rights of children to express themselves and how parents/guardians need to respect these rights. Article 24, which focuses on their medical care, states that necessary equipment and provisions should be available, when needed. Articles 36 and 37 consider a child's well-being and state that a child will not be subjected to degrading or cruel treatment (UNCHR, 1995). Thus, there are very specific and connected expectations for how to include youth in research studies.

For the case studies for this chapter, we considered the following topic areas:

- trusting relationships with researchers and participants that link to issues outside the research context;
- pharmaceutical research practices around diarrheal disease that impact infants in hospitals serving low-income families; and
- changing data collection practices during a research project so that young children can participate.

DOI: 10.4324/9781003033233-2

Case Studies and Analyses

The case studies in this chapter are based in three countries: Chile, China, and Ghana. They deal with the impact of food quality on academic achievement in young children at a boarding school (China), a pilot test of transgenic rice to reduce diarrheal disease in infants (Chile), and including youth at varying schooling levels in a study considering food availability (Ghana). Through the case studies, we see how (a) being a whistleblower can impact funding for schools, (b) regulatory expectations in different countries allow for certain studies to be conducted in some regions and not in others, and (c) limiting when children are allowed to participate in research may decrease a more holistic view of a specific topic area.

For this chapter, we provide four case study analyses due to the very interesting nature of the case study from Chile. The countries represented in these analyses are Paraguay, the U.S., Turkey, and El Salvador. For the case study analysis from Paraguay, the political and educational nuances are highlighted such that the research context is very different from that seen in China. In addition, how teachers are paid, and their work valued in Paraguay sets up a distinct situation for how a researcher would respond to the dilemma posed. For the pilot test of transgenic rice, it is clear from the case analyses from the U.S. and Turkey that while the same principles of autonomy, beneficence, and justice underpin the regulatory framework the parameters of the study would be different in all three countries examined. Finally, for the case analysis from El Salvador, it is clear that the rights of minors in research participation are well considered in legislation and ethical considerations for working with youth.

Case Study #1 (China): Maintaining a Trust Relationship between Researcher and Rural Primary School for Service Programs to "Left-behind Children" by Weihua Liu

Context

In Midwestern China, there has been migration of the workforce from rural to urban settings, where jobs are more plentiful. Many young parents who make this transition leave their children, i.e., "left-behind children," in their rural hometowns to be cared for by grandparents. Boarding service is optional, but many parents opt to have their children board at the school during the week and only return to their grandparents on the weekend. This type of oversight leaves hours of time when students are not under the supervision of either their grandparents or teachers. Local social work students noticed problems arising for many of these children, such as involvement in unhealthy peer relationships, avoiding schoolwork, and spending time in unsafe physical environments. These social work students were sponsored by a nonprofit organization to engage with rural primary schools to deploy and evaluate service interventions for primary school students, helping them to realize all-around development. The research component for this case study is about academic achievement and nutritional quality—the social work students were involved in the data collection process and evaluation of the nonprofit organization's programming.

Statement of the Issue

Teacher salaries are low in this area of the country, and schools have decreased the quality of provided meals to shift funds to increase teacher pay. Children generally do not appraise meal quality, and neither parents nor grandparents are on site to judge the quality. Social work students are on site and are aware of meal quality when they see children dump distasteful food in the dustbins. In this setting, social work students face a whistleblower's dilemma: whether to notify local governments about the diversion of funds or whether to stay silent to maintain their own, potentially beneficial presence at the school.

Outcomes

Social work students did not disclose the diversion of funds to local government authorities. They felt that government authorities were already aware of the diversion of funds, and the act of whistleblowing would not actually change the situation because there were no other funds that could be sent to the schools. Social work students also felt that the act of whistleblowing would destroy their relationship with the school, meaning that they would no longer be able to provide the services they perceived to be beneficial to students. Additionally, they would be unable to continue any ongoing evaluations of their existing projects. Instead, they have attempted to bring the issue of food waste to the attention of school administrators, hoping to engage them through informal discussion where they can highlight the relationship among meal quality, children's nutritional status, and subsequent academic achievement.

Questions to Consider

1. When a researcher accidentally discovers an unfair practice at a research location and the unfair practice is unrelated to the research purpose, how should they respond?
2. In China, social scientists may not pay much attention to ethical standards, which is related to economic, social, developmental, cultural, and political rationales. What attitude toward the violation of ethical standards is desirable for an ethics-minded researcher?

Case Study Analysis: A Perspective from Paraguay by Tsivya Jeannette Devereaux and Diana Mendieta

Context

The case study from Midwestern China emphasizes key relationships between Chinese government, families, schools, and social workers regarding ethical decisions around children's health and well-being. If a similar situation were to occur in Paraguay, how would a researcher handle an ethical dilemma that is unrelated to their research?

There are a few considerations to understand the perspective of an average Paraguayan. This case study takes place in an area where boarding schools are a popular option for children of parents who work far away all week. Boarding schools are not a well-known option in Paraguay. If they exist, they would be an option only for very wealthy people, and not affordable for parents from a rural area who must work in an urban area. However, the percentage of parents who leave their children with grandparents while working in the city is very high. Many adults tend to move to the big cities or even abroad to Argentina or Brazil looking for work. The children of these parents stay with their grandparents and/or other family members.

Like the case in China, many students in Paraguay receive meals at school. There is a meal program in public schools and children can get breakfast and lunch at school. These meals are paid for by the Paraguayan government. Traditional Paraguayan food is not generally healthy, relying on a lot of fried foods, meats, and dense carbs. There are few vegetables in the average Paraguayan diet. If a situation like the case in China were to occur in Paraguay, the quality of food would likely not be an issue. In general, parents and grandparents would be happy that their children are being fed and would expect the children to eat what is provided to them. In Paraguay, there has not been a discussion about academic performance and nutrition access.

In Paraguay, teachers' salaries include paying for classroom materials. To keep as much of their salary as possible, teachers frequently buy little to no materials for the classroom. Teachers' salaries are low, which may not be an "accepted" fact, but it is known. And in most cases, teachers decide to be teachers knowing this, but appreciating the stability, social security, retirement plan, not long working hours and plenty of vacation days. Funds would unlikely be diverted to increase teachers' salary. This issue may be different from the case in China where it seems that teachers are paid separately from what funds are given to the school. This system would also change the dynamics of the situation because it is common for teachers to pocket money that was intended for the school.

This case brings up social workers going into the boarding schools in China. In Paraguay, it would be incredibly rare for a social worker to go to a school.

Country-Specific Regulations

With the Paraguayan context in mind, how would a Paraguayan researcher go about addressing this ethical dilemma about funding and food waste that may be unrelated to their research? It is hard to say what an individual person would do, but systemically we can address what Paraguayan culture and standards would indicate. Unfortunately, Paraguay is among the most corrupt countries globally, which means people tend to do what is best for themselves and their families. This concept travels up the power chain. On a small scale, this process could mean that if a person becomes aware of an issue like the one presented, but their job and stability is on the line, they may stay quiet and not take any action.

Going along with corruption, something else to consider is the political stances of the school and the local, regional, and national government. In Paraguay, there is a strong two-party system, Liberal and Colorado. It is unfortunately common for the leaders of a party within a community to only work with other members of the same party. If, for example, the principal of the school is Colorado, but the mayor of the town is Liberal, it is unlikely for the local government to assist the school in raising funding for the student meals. If the leaders are in the same party, it is more likely for there to be collaboration between the school and government. This same principle applies between local and regional and national governments. Every step up in the government, leaders are more likely to reach an understanding and receive more funding or help if their boss is a part of the same party. If they want to do something about this unrelated issue, a researcher would be thinking about whether their political stance lines up with that of the local town.

Another variable to consider is the size of the school. If this is a large school, it would be easy for the inequities to go unnoticed or ignored. In a small school, however, there is a higher likelihood of something being changed. This topic could be related to the social pressures in Paraguayan that are heightened when there are less individuals responsible. If this is an issue with one school, it is something that could be addressed within the school itself. If it is a systemic issue, however, it would be much harder to change. Paraguayans tend to use phrases like "If God wills it, it will be" and take responsibility off the individual and accept that those are inequities that exist in the country. For example, if one or two schools of fewer than 100 students were having this issue of feeding students, a Paraguayan community may host a fundraiser or collect food donations for the students. Fundraising is a prominent activity in Paraguay since the sense of community is very strong and important. If someone is sick or needs help purchasing something out of their price range, the community may come together and organize a fair or raffle, sell food, or something similar to raise enough funds to cover whatever is needed. With this in mind, if the researcher is from the community where there is the issue, it may be easy to rally the local leaders into starting a fundraiser. However, if they are from outside the community, the researcher would feel more conflicted about a responsibility and capability of becoming involved in a "local" issue.

Addressing Ethical Issues Involving Research Participants

In Paraguayan culture, family is very important, and it comes before absolutely anything, even rules, schedules, and ethics. On one hand, there is the social worker or researcher that may try to do the right thing, since children are not being well fed, but there are many other things to consider before bringing this to the authorities. A researcher would need to ask the following questions: Would it really make a difference if I bring it up? How are politics involved in this? Do I have support to bring this up from parents, grandparents, or the whole community? Can I be sure that the authorities are going to see this as a problem and are going to do something about it? Is my job on the line? Is there a chance I can approach a group of parents and ask for their opinion and help?

Overall, if a Paraguayan researcher uncovers an issue unrelated to their research, there are many barriers to accomplishing any meaningful change. There are political issues with party alignment, there are traditional food issues where parents or grandparents may not see the food as an issue, and there are community belonging issues of responsibility as an outsider looking in on a community. Thus, the whistleblower's dilemma would not apply for this research context.

Case Study #2 (Chile): Experimenting with Pharmaceutical Transgenic Plants for Medical Treatment for Infants: Social, Legal and Ethical Issues by Eduardo Rodríguez and Fernando Lolas

Context

Every year, two million children around the world die of complications from diarrhea. The condition is particularly serious in developing countries. Even when not resulting in death, diarrhea is associated with malnutrition and weakening of the immune system. Research has shown that breast-fed children have a much lower incidence of diarrhea.

A U.S. bioscience company and a U.S. university in the western part of the country planned a study with lactating infants to evaluate the effectiveness of a pharmaceutical transgenic rice for use with acute diarrhea. This transgenic rice was modified to produce two key protecting proteins found in mother´s milk: lactoferrin and lysozyme. Researchers planned to add these proteins to oral rehydration solutions to reduce the severity of diarrhea. Researchers thought that if found effective, it would be an affordable solution against diarrhea in low resource settings. The experiment was performed in 2004 in public hospitals in Lima and Trujillo, Peru with the approval of the Peruvian Ministry of Health.

The Nutritional Research Institute, which performed the experiment in Peru on behalf of the U.S. stakeholders, stated that all national and international norms were respected with regard to human subject research and bioethics in this study.

Statement of the Issue

The rice, grown in Chile, was approved by the Peruvian Agricultural and Farming Services. Both the U.S. and Peru prohibit the growth of this rice in their own countries. However, Chile allows the production of transgenic crops for export and supplied the rice for this study. One concern about using pharmaceutical transgenic crops is the potential to contaminate food, with medicines or chemicals, which is destined for consumption. Culturally, transgenic products are looked upon with suspicion in Peru.

The experiment was performed in public hospitals that serve poor populations. Peruvian lawyers considered this intervention a violation of children rights, since it disregarded the primacy of the Superior Interest Principle that one must do what is most beneficial for the child. In this case, outcomes were not known for the solution that contained the lactoferrin and lysozyme produced in transgenic rice, and benefit could not be determined. Ethically, all safeguards must be taken in place to avoid health risks especially in vulnerable populations, such as poor infants.

Outcomes

The experiment involved 140 hospitalized infant boys, five to 33 months old, who received one of three oral rehydration solution (ORS) formulations: low osmolarity ORS formulation (the WHO-approved formulation), rice-based ORS, or rice-based ORS fortified with lactoferrin, and lysozyme produced by transgenic rice. Some of the children who received fortified rehydration solutions experienced strong diarrhea and other severe allergic reactions.

Questions to Consider

1. How should the Superior Interest Principle presented by the lawyers be addressed when it differs from the principle of equipoise for research?
2. What social issues should be addressed in international studies such as the one presented? Consider issues of transgenic rice, the balance of power and burdens in the international collaboration, etc.
3. How might you address the ethical principle of justice as it relates to (a) enrollment only of boys and (b) enrollment only of low socio-economic status infants?
4. In what other way(s) do you think this study could have been performed?

Case Study Analysis: A Perspective from the U.S. by Jeffrey R. Botkin

Context

The case highlights several complex regulatory, ethical, and social issues in the conduct of research with children. In this case study analysis, I will highlight the regulatory issues for projects conducted under the auspices of the U.S. and briefly discuss several ethical and social considerations. My general conclusions are that this project is likely to be considered appropriate under regulatory and ethical standards in the U.S.

A basic consideration at the outset is whether a proposed project addresses an important health concern for young children. Further, we should consider whether the research can be conducted with adults or older children who can provide informed consent. In this case, diarrhea is a major cause of morbidity and mortality in infants and young children in low-resource settings around the globe. We know that oral rehydration solutions can mitigate the impact of this condition in a cost-effective way. In the context of this study, it would be important for the investigators to articulate why current approaches to oral rehydration (i.e., the WHO formulation or the rice-based solution) are not fully adequate and why the proposed experimental formulation has reasonable prospects for safety and efficacy. If there are animal models for infant diarrhea, investigators should present data on the efficacy and safety of the fortified solution in the animal model. Further, investigators should consider conducting a Phase I and Phase II trial of the solution before moving to a larger Phase III study. In this case, it is reasonable to conclude that the project addresses a serious health problem in children, that the proposed intervention has prospects of success, and, because the issue is infantile diarrhea, the research must be carried out with young children.

A critical component of the ethical justification for this research is the concept of equipoise. Equipoise is a complex concept, but the core idea is that investigators and clinicians must not already know the answer to the questions being posed in the research. If the health community can draw a firm, evidence-based conclusion that the WHO formulation is fully adequate as a treatment or is more effective and safer than the experimental ORS, then conducting additional research on the question would not be ethically appropriate. However, there are a number of complexities in deciding what the health community already "knows" in this regard. Often several trials must be conducted before firm conditions can be drawn about the safety and efficacy of any intervention. Further, we might accept a lower level of safety in some circumstances for a higher level of efficacy, or vice versa. There are also often disagreements in the health community about what we "know" to be true; that is, some members of the community may be convinced of a particular conclusion by existing data while others are not. The point is that those responsible for designing, reviewing, and conducting the trial must be able to demonstrate equipoise; that is, there must be legitimate uncertainty about whether the fortified ORS will prove to be more effective and safer compared to existing approaches.

In this case, the investigators are comparing the experimental intervention against two other ORS formulations. Both of these alternatives are in common use for infant diarrhea and can be considered standards of care. Notably, the investigators did not propose to compare the fortified solution against no oral rehydration; that is, a placebo control. This comparison might make for a stronger experimental design, particularly if there is any question about the overall efficacy of ORS for infantile diarrhea. However, if we know that no ORS is inferior to the commonly used ORS approaches, denying children in the study a standard of care treatment for a serious condition would be unethical. In this context, the Superior Interest Principle as described in the case would be relevant. That is, it is unethical to conduct a trial in which you know children are being placed at increased risk compared to available, evidence-based alternatives. However, I would claim that the Superior Interest Principle (or "best interest standard" in common U.S. parlance) is not violated if the experimental intervention has reasonable prospects of benefiting the children in the study and none of the study children are receiving less than standard-of-care interventions.

Country-Specific Regulations

In the U.S. regulations governing research with children (45CFR46 Subpart D) (2), there are four categories under which research can be approved (USDHHS, 2020). In the first, the research must be considered minimal risk, meaning the risk is no greater than the risks of everyday life. This category is almost never appropriate for studies involving a new drug or internal device. The second category (45CFR46.405) is most relevant to this case and to most other cases involving drug or device interventions. Under this category, the risks can be greater than minimal if there are prospects of direct benefit to the research participants. The risk must be justified by the anticipated benefits, the benefit/risk ratio must be at least as beneficial as alternative approaches and, of course, parental permission and child assent must be obtained, when appropriate. Under this category, the intervention can confer a substantial risk if there are prospects of a proportionate benefit.

An underappreciated complexity of these regulatory categories is that different categories can be appropriate for different groups within the same study. For example, if a study has a randomized, controlled design, participants will be assigned either to an intervention group or to a control group. For the intervention group, the regulatory category involving a prospect of direct benefit would be appropriate for consideration. However, the control group does not have a prospect of direct benefit, meaning this category is not relevant. The control group should be considered under the minimal risk category. This approach reinforces the notion that control groups generally should not be placed at more than minimal risk when there is no prospect of benefit.

These considerations suggest that the proposed research in this case is ethically designed and approvable under U.S. regulations, assuming the investigators can effectively argue that the control groups are receiving standard of care approaches and that the experimental intervention has reasonable prospects of benefiting children with this serious condition. A review of the study publication indicates that the experimental ORS led to a meaningful benefit for the children in the intervention group (Leighton, 2006; Zavaleta et al., 2007). The occurrence of some serious adverse effects in an intervention group does not undermine a pre-study determination that the trial was ethical. There is almost always a risk in research that the experimental intervention will prove to be less safe or less effective than existing alternatives.

Addressing Ethical Issues Involving Research Participants

The case also raises some interesting issues of justice and social standards. The fact that the research was conducted in hospitals serving low-resource communities could raise concerns about justice. There is an unfortunate tradition of poor, marginalized, and institutionalized children being used for research for matters of convenience and cost for investigators, not because the research would answer questions particularly relevant to the health of those children. If a new drug will not be used to address the future needs of a low-resource community, then it would be unethical to recruit such participants for research simply to achieve lower study cost or greater convenience. In this case, infantile diarrhea is more common in poor communities and the experimental ORS is likely to be a widely used and cost-effective tool in such communities. Accordingly, recruiting participants from environments where the problem is most prevalent, and the research questions are most salient, can be justified.

The decision to enroll only boys in the study is not explicitly justified in the case study or in the publication. Because infantile diarrhea is a serious issue for both boys and girls and there is no apparent a priori reason to think that ORS would function differently in girls than boys, this decision is ethically questionable. An ethics review committee for this study should specifically address this question and should insist that both sexes be included if there is no justification based on anticipated safety or efficacy.

The other interesting element in the study is the use of a transgenic product to create the fortified ORS. Transgenic agricultural products have been controversial in many areas of the world and are subject to specific regulations and consumer notification practices. In this case, Peru prohibits the growth of the rice product in their country but not the use of the product. From my perspective, there is little scientific justification for the broad safety concerns about the safety or efficacy of transgenic crops (environmental issues aside). However, specific modifications such as the addition of lactoferrin and lysozyme, as in this case, should be carefully studied to assess their safety and efficacy. This procedure is, of course, the purpose of the study. Yet community concerns about issues like the use of transgenic crops should not be summarily dismissed by scientists even if those concerns lack much scientific validity. The scientific enterprise is dependent on public trust and public support. Simply dismissing public concerns could be interpreted as investigators concealing important aspects of a study. In cases such as this, it would be wise to engage the public during the design of the study to (a) ascertain public sensitivities about this issue and (b) design informational and consent language that effectively informs potential participants about this element of the study.

Case Study Analysis: A Perspective from Turkey by Muhammet Arslan and Nüket Örnek Büken

Context

Diarrhea stands as the second leading cause of death in children under five years old. Being usually a symptom of an infection in the intestinal tract, diarrhea causes severe dehydration and loss of fluid. If left untreated, along with poor nourishment, it eventually leads to death. Diarrheal diseases mostly result from contaminated food and water sources. Nearly 780 million people in the world have trouble accessing clean drinking water, and 2.5 billion lack proper sanitation conditions (World Health Organization, 2017).

Diarrhea can be prevented if access to "safe drinking water improved sanitation, hand washing, exclusive breastfeeding for the first six months of life, personal hygiene, food hygiene, education and rotavirus vaccination" are provided. In terms of treatment, rehydration with oral rehydration solutions (ORSs), Zinc supplements, and nutrient rich foods are widely used and shown to be effective (World Health Organization, 2017).

Country-Specific Regulations

Turkey carries huge potential for clinical trials due to its cosmopolitan base of ethnicity, disease burden, and rarity of genetic diseases. As noted by researchers, "as in line with EU [European Union] directives, Turkey strictly follows international documents and bindings" (Atıcı, 2009; Büken, 2017; Örnek Büken & Küreci, 2018; TİTCK, 2015). Basic liberties and human rights are protected and held at a higher value than the possible outcomes of a study. Clinical trials in Turkey follow Good Clinical Practices (ICH GCP) standards, directives of the European Union, Declaration of Helsinki, Oviedo Convention on Human Rights and Biomedicine, and other related documents. Each study needs to obtain the approval of an independent ethics committee and of the Turkish Medicines and Medical Devices Agency (TİTCK) as well as the informed consent of the participants (TİTCK, 2015).

In case of children, Turkish legislation requires the consent of the parents/ representatives with attention to the minor's approval in participation. The Declaration of Helsinki (Art. 28, 29) (World Medical Association, 2013), Oviedo Convention on Human Rights and Biomedicine (Art. 16, 17) (Council of Europe, 1999), Law on the Practice of Medicine and Medical Sciences (Art. 70) (Official Gazette, 1928), Regulation on Clinical Trials of Drugs and Biological Products (Art. 5, 6) (TİTCK, 2015), Regulation on Patient Rights (Art. 24, 26, 35) (Official Gazette, 1998), Turkish Penalty Law (Art. 90) (Official Gazette, 2004), The Basic Law on Health Care (Additional Art. 10) (Official Gazette, 1987) include statements and requirements concerning the participation of children to a clinical study. According to the Regulation on Clinical Trials of Drugs and Biological Products (Turkish Ministry of Health, 2013) (Art. 5), clinical studies involving children also need the approval of an expert pediatrician.

The production and exportation of GM plants and animal products is unauthorized in Turkey (Hayırlıdağ et al., 2016). Relevant matters are regulated in Biosafety Law by the Ministry of Agriculture and Forestry (Official Gazette, 2010). A National Biosafety Committee was established to monitor and control GM product activities. The use of GM products in baby foods, formulas and supplementary foods is forbidden. Therefore, the presented product does not have an application for Turkey as the use of GM products in infant formulas are not permitted. Turkey has signed the Cartagena Protocol on Biosafety, which regulates matters on transgenic products (Kıvılcım, 2012). The principle of precaution embedded in the protocol gives the state the right to ban or limit GM activities (Kıvılcım, 2012). There is also a general attitude in Turkey to reject GM products due to some cultural reasons and the negative representation of GM products in the media (Taş, Balcı, Yüksel, & Şahin Yeşilbuçuk, 2015). There is an ongoing and growing approach among parents to feed their children with organic and natural components (Eti İçli, Anıl, & Kılıç, 2016). In that sense, even if a GM product was properly tested and proven effective, there might be some conjunctural objections. Considering the prevention and treatment options for diarrhea available in Turkey where access to clean water and food is not a serious problem (Aküzüm, Çakmak, & Gökalp, 2010), proven and cost-effective alternatives would restrain such a GM product from finding a place in the market.

A researcher in Turkey ideally would not participate in such a study before all the safety tests were completed. As GM products in infant formulas are forbidden, such a study would not be legal to initiate. Besides, the study in the presented case would not be allowed on legal grounds. Ethical committees would not approve the case as it might violate the integrity of scientific research, ethical principles, minors' best interests, parents' rights, and the interests of future generations.

Addressing Ethical Issues Involving Research Participants

Clinical research is an essential part of medicine, which carries the same motivations and importance as clinical practice. Considering that children are different than adults in many ways and they cannot be regarded as mini-adults, it is important to develop age- and physiology-related prevention and treatment options for them (Büken, 2017). Clinical research on children should aim at benefiting children individually, as well as benefiting the pediatric population as a whole; it should also aim to protect the children from harm (Örnek Büken & Küreci, 2018). Minors are considered a vulnerable group due to their lack of ability to consent. Additional vulnerability might result from disabilities, and/or environmental and social conditions. Meticulous attention should be paid for the protection of children, and to any other vulnerable group (women, racial or ethnic minorities, elderly people, prisoners, chronically ill patients, migrants, etc.), especially in case of an invasive intervention like clinical research (Davidson & O'Brien, 2009).

As stated in Article 8 of the Universal Declaration on Bioethics and Human Rights, "in applying and advancing scientific knowledge, medical practice and associated technologies, human vulnerability should be taken into account" (UNESCO, 2013, p. 5). Personal integrity of individuals and groups with special vulnerabilities should be respected and protected (UNESCO, 2013). Researchers and research protocols should guarantee that any study involving children protect children's rights and interests. Research on children should only be performed when necessary and beneficial to the child. Ideally, the risk should be minimal. In the case of life-threatening conditions and the absence of alternative therapies, more risks could be taken after careful consideration of the costs and benefits (Modi et al., 2014).

Since children (i.e., the infants referenced in the study) are unable to consent, the term *assent* is used to describe a child's voluntary affirmative agreement in participation to a study. A formal consent must be obtained from the parents or legal representatives on behalf of the child, together with the child's own assent, if possible. Children and their parents or legal guardians should be informed that they have the right to exclude themselves from the study at any stage, without any compensation or loss of therapeutic attention (Büken, 2017; Modi et al., 2014).

Clinical research ethics committees play a vital role in the management of clinical trials and of ethical and legal concerns. They serve the best interest of the child by weighing the costs and benefits. Ethics committees also warrant the integrity of research, protecting researchers, funders, and all relevant subjects (Modi et al., 2014). The principles for these ethics reviews are respect for autonomy, nonmaleficence, beneficence, and justice. The impact of these principles on research with infants is discussed below.

Respect for Autonomy

The principle of autonomy recognizes an individual's integrity and independent decision making. It ensures that participants in clinical research are able to make their own informed decisions. In the example of the case above, autonomy becomes highly important especially when participants are vulnerable. Considering that the clinical intervention is performed in public hospitals that serve poor populations, certain individuals become more open to exploitation due to their lack of access to basic medical needs. Parents in these situations may be in serious need of basic medical care, where they see the intervention as their only chance to access treatment. Hence, they might feel a sense of parental burden to participate in the study. Since children at a very young age cannot consent or assent, their parents/representatives should be fully informed, and the necessary conditions for them to make independent decisions must be assured. In extreme cases like this, both parents' mutual decision making could be encouraged instead of only one parent's decision. The present case raises the question of whether the parents or guardians of the children were sufficiently informed to provide informed consent in terms of the issues above. Considering that the consent being "informed," obtaining informed consent should not mean only to have a "written" document but provision of all elements: explanation, comprehension, competency, and voluntariness.

There might be situations where the responsible person for recruiting is held in high esteem and their decisions, actions, or intent cannot be objected. This issue may be more common when people are in serious need of basic medical care. Therefore, parents might not find the courage to raise their concerns as they might think that their objection would cause them further sanctions by the people/groups in charge. The same concern is pertinent in situations where for example a clan leader decides for the actions of their people.

Nonmaleficence and Beneficence

Clinical studies must serve the best interests of the participants. Meticulous analysis is needed in this respect, considering the vagueness of the effects of transgenic products. The proposed transgenic product has unknown aspects such as potential to contaminate food where it is grown and containment of human genes with the risk of unknown interactions (FoE Africa, 2007; Hayırlıdağ, Arslan, & Örnek Büken, 2016). The Nuffield Council on Bioethics has published a report about the ethical issues arising from the use of genetically modified (GM) products. The report states that the GM products may have effects on public health, environment, traditional agriculture, domination of industry, and divergence from naturality (Nuffield Council on Bioethics, 2003). Therein, it is fair to say that what is possible to do is not always permissible, and ethics determine the limit in between (Hayırlıdağ et al., 2016). Any study planning to trial a transgenic product on humans should conduct animal tests first to determine the safety and efficacy of the product. The case does not provide any information on whether animal tests were done before the clinical intervention (Freese, 2007).

Justice

The principle of justice plays a vital role in creating balance within the society and preventing any inequalities. As Rawlsian perspective emphasizes that justice should mean preserving the rights and interests of the least advantaged while providing the equal distribution of opportunities and services (Rawls, 1999).

The case at hand leads to concerns about the application of untested products in economically challenged societies, where people lack access to basic medical care and therefore might feel forced to participate. The principle of justice highlights that people's vulnerabilities should not be exploited for scientific advancement. Vulnerabilities should not be seen as an opportunity to initiate a trial, and trials should not be regarded as mere "procedures" to gain a place in the market. It is also questionable whether the design of this study would be permissible, were it to be conducted in a high-income country.

The study only involves male children, which is another ethical issue considering that it creates an unjust situation both in case of unknown benefits and harms. According to studies on provider and caregiver preferences of gender for treatment of infants, girls, especially in poor countries, receive less prescription and treatment, and are less often brought to the hospitals (Yount, 2003). There are places where girls are discriminated against or neglected through the allocation of food and healthcare, even by their parents. The same biased preference was also observed for the treatment of diarrhea among boys and girls (Paredes, de la Peña, Flores-Guerra, Diaz, & Trostle, 1996; Yount, 2003). Recruitment only of boys or girls should only be justified when sex selection is required by study design and backed up by scientific facts. Without justification, selection only of one particular sex is against research ethics (Örnek Büken, 2018).

Additionally, the proposed treatment for diarrhea might be a distraction from existing and proven solutions like the provision of clean water, sanitation, ORSs, and breastmilk. Therein, the question of pricing also arises. Considering that there are people in the world whose monthly income is equal to a cup of coffee in some places, some might not be able to afford extra costs. Even though it is shown to be effective and safe, the product will have to have a price tag on it and this new costly product might not be available for everyone in the community. There emerges another unjust situation where people were exposed to risks but do not derive any benefit. The existence of alternative preventive and curative options and the pricing requirements raise concerns about the purchasing power of the targeted population. Thus, it is ambiguous whether the study has any realistic goals to reduce the effects of diarrhea in poor populations.

Case Study #3: Assessing the Accuracy of Data in School-aged Children in Rural Ghana (Ghana) by Rebeca A. Michael and Lydia Debrah

Context

In 2019, a cross-cultural research team working in Ghana, West Africa used dietary surveys to obtain data about the type and amount of food that youth in a rural community ate. Since specific types of dishes and foods are unique to this region of the world, there is not a standard and/or validated dietary survey available to determine the nutritional value of food youth eat. This baseline study was aimed at determining if a population of children in rural Ghana were able to recall what they had eaten the day before answering survey questions. Participants were students in a public school, age range eight to 16 years old, who were willing to participate in the survey process. Verbal assent was obtained from all participants.

Statement of the Issue

In this rural context and given the English literacy level of the youth, researchers were looking for a creative and informative way to collect the dietary data from youth. Students in this area of Ghana typically speak one or more non-English languages at home. Even though it is the language used at school, English is a less comfortable language for most students. Because of the many spoken languages, the research team decided to conduct the study in English, rather than attempting to translate into multiple languages. Teachers, and some parents, felt that students would not be able to complete this activity with instructions provided in verbal English. Researchers were concerned about the potential for inaccurate recall for a variety of reasons. First, there was a concern that students would not understand instructions and might not accurately report their consumption of food and drink at meals and/or between meals. There was also the potential for variation in the amount of probing to help recall consumption during the past 24 hours, as teachers stepped in to help some of the younger children in a pilot test of the dietary data collection. One final concern was the potential for social desirability bias, as a result of students wanting to participate in the data collection, even if they could not accurately recall what they had eaten the day before. As a result, the research team decided to collect data using drawings, as this was seen as a way to minimize difficulty in English language understanding, as well as decrease the need for children to require help from teachers.

Outcomes

Children were willing to draw pictures of their meals, and the descriptions were generally apparent to the researchers. This creative approach to collecting a 24-hour dietary recall was made possible because the teachers, who were very supportive of the project overall, provided instruction in the local languages spoken by students. Older children were able to both draw what they had eaten and write the names of the items. Younger children could draw what they had eaten, and then dictate the names of the items to an adult. This innovative data collection resulted in a broad description of foods eaten by children that matched up well with responses from adults in a separate data collection activity.

Questions to Consider

1. How should researchers inform themselves about conducting research with minors?
2. How can researchers ensure that the data collected are as accurate as possible?

3. How can adults in the research setting provide assistance without influencing the information obtained for the study?
4. What concerns might there be about collecting verbal assent from children in a group setting?
5. What are the advantages and/or disadvantages of conducting research with minors in cross-cultural settings?

Case Study Analysis: A Perspective from El Salvador by Jorge E. Lemus

Context

The responsibility for ethical behavior in scientific research in El Salvador falls mainly on each researcher, who applies the ethical standards established in their area of specialization. Institutions of higher education and research centers and institutes have their own guidelines and ethical considerations for research, which are generally supervised by a research committee or council. At the national level, the only regulating committee is the National Health Research Ethics Committee (CNIS) in the medical and biological sciences (CNIS, 2021).

Country-Specific Regulations

There is a long history of abuse and discrimination against the Indigenous communities in El Salvador. At present, that situation has changed, but most Indigenous communities are still distrustful of outsiders and, therefore, reluctant to participate in any project. Thus, the first step for any researcher is to gain the trust of the community. Fortunately, public schools are trusted institutions, so approaching the schoolboard and explaining to the parents the scope of the research in plain language is usually enough to obtain authorization to conduct any research within a school.

Besides school permission, any research or activity involving children is subject to the Law for the Integral Protection of Children and Adolescents (LEPINA), which requires a written agreement from the parents or legal guardians of the children involved in the research. Research like the type described in the Ghanaian case would also require a written authorization from the regional agency of the Ministry of Education. That is, for any school children to participate in a research project, a written authorization from the schoolboard and the Ministry of Education must be obtained.

A major concern with research of this kind is how to obtain the informed consent from the participants (the children) themselves. Article 46 of the LEPINA clearly states the children have a right to privacy. Asking about eating habits could be regarded as an invasion of their privacy. For this reason, informed consent is paramount in surveys that ask about private practices. When children are too young to comprehend the nature of their participation in a survey, their parents or legal guardians are responsible for their participation. However, if during the research, a child is not willing to participate, the researcher must not compel them to do so, even if their guardian agrees to the research.

In a school setting, the willingness of the children to participate in the survey may be influenced by when the survey is provided. If given during playtime, most children would be reluctant to participate, regardless of the approval of their parents. In the case study, there is no specification regarding the time the survey was given. Article 90 of the LEPINA establishes the right children have to leisure time, sports, and play. Using that time for conducting research may be regarded as a violation to their rights. Therefore, the researcher must come to an agreement with teachers and guardians to conduct the research at times where there will be no interference with their playtime or any other leisure activities. Passing the survey during regular classes could also be regarded as invasive and limiting to the children's normal learning according to the school curriculum.

Addressing Ethical Issues Involving Research Participants

For research involving younger children who do not yet know how to read or write, the use of pictures, drawings, and videos is advisable. For children from birth to four years old, the parents must be present during the survey and many times interact with their children. For example, in research conducted by the author to measure the cognitive and psychomotor development of Indigenous children attending the Cuna Nahuat (an early Nahuat-language immersion program), the questionnaire was directed to both parents and children. We asked parents about the behavior of children at home, and we asked the children to carry out different tasks by themselves, without the help of their parents. This research could not have been done without the presence of the parents.

Nevertheless, not understanding the language or the cultural practices of the community may lead to conflict with parents or elders who might find the researcher's approach culturally invasive. As already mentioned, Indigenous communities are distrustful of strangers. The language barrier will always come between the researcher and the research subjects. Linguists are used to facing this situation at the beginning of our research projects in communities whose first language is different from our own. The use of images is always useful, both with adults and children. However, it is recommended to use the language of the study subjects, even at a basic level, to establish a level of trust and empathy with the project participants. This use of the language is important, regardless of the presence of an interpreter. Hence, when the researcher has no command of the language, the use of pictures and drawings is highly advisable, as in the Ghanaian case. The only problem arising from this approach is when children are too young to do their drawings, in which case using pictures would be the only choice.

Once they agree to help with the project, parents and teachers can create a comfortable environment for children to participate in the research. Parental involvement in the project can be important to obtain informed consent from the children. The project's objectives must, first, be fully understood by parents or guardians. Then, they can help translate, in a language the child would easily understand, the purpose of the research. The use of visual aids to make sure younger children understand what is expected of them in the project and what it is for is always advisable. Only in this way, we can be sure that the children are informed and, therefore, capable of giving their consent to participate. In my experience, sometimes children agree to participate in a survey because they think it is cool and all their friends are participating; however, when they start answering the survey or participating in the research activities, they may find it too difficult or boring and would like to withdraw from the project. When this happens, we must allow them not to continue in the process and thank them for their time. If we have brought presents for the children as a reward for their participation, we should give them to all, regardless of their participation. Parents should be informed about their children's withdrawal from the project and asked them not to question or reprimand their children for their withdrawal.

Finally, something I find difficult to do at the end of the research with children is sharing the results of the research with them. In the Ghanaian case, there is no evidence that this was done. The researchers must be highly creative in presenting the results of the research to the children. All technicalities should be avoided, and the use of visual aids must tell a story understandable by children. A different presentation should be prepared for parents, teachers, and community leaders.

Summary of Chapter

In summary, research relating to children requires special attention, both regarding regulatory practices and implementation. Due to the vulnerability of minors, lack of full understanding of the research, and potentially being subject to coercion, researchers need to be able to account for or address these issues and more when proposing their research. The case analysis from Paraguay raised political and cultural issues when conducting research and how differing views can help or hinder the study progress based on the affiliations of the researcher. Community is a core aspect of the culture and communities are less likely to interact or participate if it is an outsider doing the asking or approaching. Gaining community buy-in is a large part of research, especially when dealing with children as the parents and often the schools themselves also need to be aligned with the same view of the issue at hand.

Transgenic rice or genetically modified (GM) products raised both regulatory and justice issues. Even though this type of study might be accepted in the U.S., it brings the question as to why go straight to an experimental formulation versus oral rehydration in this instance. Since not much is known about experimental products, it would be unethical to place study participants (i.e., children) at an increased risk as compared to the readily available evidence-based alternatives. However, this also leads to what is readily available in the country at hand. GM products are not authorized in Turkey and are prohibited in children's products such as formula. Clean water and sanitation are also readily available and thus do not present the same issue as what is seen in Chile. Justice issues brought up by both the U.S. and Turkey highlight research conducted on poor, marginalized populations. It is unethical if this population is used for the sake of convenience or to reduce costs. These populations should only be considered if they are the same populations that will benefit from the outcome. Even still, due to a negative history with research, every effort should be made to ensure that these poor, marginalized populations do not feel forced or coerced to participate. GM products are also controversial and tying into the previous case study analysis, community buy-in is needed for this type of research to be accepted. Since parental consent is needed, researchers must be able to gain trust and not act as if they are trying to conceal results or aspects of the research. Cultural reasons could also be hindering research if the general view of the product is negative, and the general attitude is to reject the product. Sharing study results could be one method of including community in developing appropriate outcomes based on the research.

Responsibility to ensure the research is conducted in an inclusive and ethical manner was another important issue raised. As compared to the Ghanaian case study, in El Salvador not only does the responsibility of the research fall to the researcher, but it also falls on the school if that is where the research is being conducted. The rights of the child come first, and research is not allowed to interfere with their education and other activities. In this instance, not understanding language and cultural practices could lead to conflict with the parents or elders. This challenge is why it is important to have community involvement to help create a comfortable and respectable environment. If the parents or elders are not comfortable with the research or issues being studied, the children would not have any trust to participate. In addition, if the parent or elder who consented was okay with the implementation, but the child is not, the researcher must have the responsibility to not force or continue the research and give autonomy to the child.

This chapter summarizes the importance of having regulations and responsibilities in place to provide ethically sound research when involving children. However, the researcher must take the responsibility and initiative to adhere their research with the rules and regulations in place within their country as this shows how a study can differ based on different cultural designations.

Chapter Exercises

1. Consider how minors are involved in research in your country or community. What are the regulations or community expectations for assent, parental permission, or consent? Make a list and compare to the information in this chapter.
2. Identify the major ethical concerns in your country or community for minors to enroll in research? How would you handle this for your own research, if you wanted to work with minors?
3. How would you develop a results-sharing discussion with parents in a community? What are the most important elements to this activity?
4. Write your own case study about a research study with minors. Share with your peers. What elements did you address in your case study that your peers would not consider a concern?

3 Female Participants

Women make up half of the world's population and yet research has mostly been conducted with and included men (Mastroianni et al., 1994). It was originally thought that men and women were biologically similar, even though texts and science would often negate female bodies by using different verbiage to describe their functioning (Martin, 2001). Current research has concluded that women and men are affected differently by the same illnesses and have different reactions to the same treatments and medications. For research conducted with male participants, study investigators have assumed that there will be the same outcomes when treatments and/or medications are applied to women; however, this finding has not been the case for many studies and interventions. According to the American College of Obstetricians and Gynecologists, it is important to include women in research studies to ensure valid and generalizable conclusions about health status, course of illnesses, and impact of treatment for female participants (ACOG, 2015). In March 1994, the U.S. National Institutes of Health (NIH) published new guidelines requiring the inclusion of women as subjects in clinical, biomedical, and behavioral research efforts (1994). These guidelines also suggested the initiation of programs and support for outreach efforts to recruit women into clinical studies. Even though efforts have been made for research to broaden the inclusion criteria for studies, there are still many that have not yet been opened to include women around the globe.

Women themselves are not considered a vulnerable population but women who are pregnant are considered for protection. Pregnancy adds to the complications when considering women as research participants, because not only does the welfare of the mother need to be considered, so does the welfare of the unborn fetus. Women who are pregnant may react to interventions differently than women who are not pregnant. Complications can arise, especially in dealing with drug interventions when the drug may have been approved for usage on women, but there is little to no research on its effects on the fetus.

Ethical complications may arise for women as research participants, depending on the culture in which the research is taking place. In some cultures, women do not have the authority to make the final decision when it comes to their health care and/

DOI: 10.4324/9781003033233-3

or to participate in research studies. Thus, research studies may not have a truly representative sample and will not be able to generalize findings across the sexes. This lack of inclusion can create biases and differences among cultural groups, increasing a lack full and well-grounded data (Mastroianni et al., 1994).

Researchers have also typically been men. This gendered view creates bias or a lack of understanding when deciding on and designing research geared toward women (Dumon, 2017). Male bias is when male perspectives are taken up by the readers or observers of that research. A male norm is perceiving the male expression to certain experiences as the standard and seeing anything else as less or an outlier (Mastroianni et al., 1994). By having women conducting and participating in research, doors will be opened for a more comprehensive understanding of research parameters and/or health needs for understudied groups. For this chapter, we are focusing on four areas about including female research participants:

- Sex-focused language in research
- Pregnant participants
- Best practices to recruit women into research given collective ideologies in communities
- Impact of research targeting women

Case Studies and Analyses

Four countries (Ghana, India, Polynesia, and the U.S.) are represented through these case studies. The case studies in this chapter deal with the impact of the cultural environment on women's research understanding and participation, how screening tests may change participant behavior, and how participation in a study may endanger participants. These case studies demonstrate how (a) women's participation in social and behavioral studies may impact other family members due to many traditional and cultural expectations and roles that women fulfill in their households, (b) unintended consequences of providing prompt and accurate health information can lead to dangerous decision making, and (c) research into domestic violence may contribute to an increased risk of abuse in the short term.

The case study analyses for the three case studies presented here come from three country perspectives (India, Morocco, and Pakistan). The case study analysis from Morocco focuses on the expansive role of women with respect to their household, and that understanding the cultural and traditional roles women play in both urban and rural settings may impact their effective participation in research. The case study analysis from India discusses potential limitations of institutional review board (IRB) oversight in communities with little experience with ethical decision support and oversight and how such limitations complicate informed consent for individuals. Finally, the case study analysis from Pakistan focuses on the complex definition of domestic violence and the implications of discussing domestic violence in family and/or public settings.

Case Study #1 (Polynesia and United States): Research with Conflicting Obligations: Research Norms versus Cultural, Social, and Environmental Norms by O. Fahina Tavake-Pasi and Brenda Ralls

Context

With the help of Community Faces of Utah (CFU), the National Tongan American Society along with partners from the African American, African, American Indian, Latino communities, the Utah Department of Health and University of Utah worked together to address the health of women in these five ethnic communities (Tavake-Pasi et al., 2016). Wellness coaches were hired and trained to address the physical activity and eating behaviors of women in these communities. Women from each community were hired and trained as Wellness Coaches. Each coach then recruited women from their own communities to participate in a 12-month study. The Wellness Coaches used motivational interviewing techniques to help women set behavior-change goals. Data were collected at baseline, as well as at four, eight, and 12 months, on self-reported health behaviors and measured anthropometrics. As data from the Pacific Islander women were gathered, it was surprising that every single one of them reported not only their personal data but also reported on the changes in the eating and physical activity behavior of their children. The reported results clearly showed a decline in risky health behaviors and conditions among study participants. Marked improvements in health behaviors were also noted for participants' children.

Statement of the Issue

Originally, the study had an "individualistic" approach that used one-on-one coaching and focused only on the health of the recruited women. As the project started and Wellness Coaches gathered critical information, it was obvious that certain issues and contradictions between the original boundaries and "perceived norms" of the study were being stretched. The study did not include the realities of the cultural, social, and environmental norms and interests practiced in the women's home and community settings. Researchers took for granted that institutional and research norms were going to be the practices followed by the participants. In the Tongan community, there's a saying, *Ko 'eku mo'ui ko 'eku fanaū* (My life is my children). Parents and grandparents value and live to help their children from birth to death. As this research project had a healthy goal objective, in the minds of Pacific Islander women, it would be normal and obvious to include their children in this project. So, they did! Not only did the study improve the nutritional value of the children's diets, but the children participated in physical activity opportunities with their mothers.

Outcomes

When women started reporting their activities and outcomes, it was apparent that the results of the project had an unplanned "rippling effect" that positively affected the participants' children. At 12 months, over half of the participants reported improvements not only for themselves, but in the eating and exercise behaviors of their children. Seventy-two percent of the participants reported that their children's participation encouraged them to engage in healthy behaviors. The reach of the study was changed to include children of participants. It was recommended at the end of this study that "future research should further evaluate the broader family reach of impact of programs targeting women" (Tavake-Pasi et al., 2016).

Questions to Consider

1. How were the cultural, social, and environmental factors—especially the complex interplay of interests—taken into consideration in the research process, activities, and IRB expectations?
2. How might the analysis need to change, once social-cultural norms were accounted for in this study?
3. As a researcher doing research in an ethnic community which may have different social and cultural norms, how will you work with conflicting obligations such as following the required IRB approval or assessing the initial outcomes, while on the other hand abiding by the social-cultural norms for ethical behaviors in the participants' home and community contexts?

Case Study Analysis: A Perspective from Morocco by Doris Gray

Context

Morocco is a Muslim-majority country, located on the northernmost tip of Africa. It is separated from Spain by only nine miles at the Straits of Gibraltar. This geographical closeness to Europe, plus the fact that within Moroccan territory there are two Spanish enclaves of Ceuta and Melilla, makes Morocco one of the countries in Africa that is geographically, historically, and culturally most closely connected to Europe. The population is primarily Arab and Imazighen, often called "Berbers." A Moroccan proverb states: "Scratch a Moroccan and you find a Berber." Moroccan Arabic (*Darija*) is the national language along with several Amazigh languages, while in urban areas French is still widely used owing to its history of being a French protectorate from 1912 to 1956. Though primary school enrollment is 97 percent, life expectancy 76.5, and fertility rate 2.4 (World Bank Country Reports, n.d.), these figures do not offer a comprehensive picture of Morocco. In rural areas, illiteracy among women is more than 60 percent and school dropout rates are high yet not officially documented (World Bank Country Reports, n.d.). Morocco is a constitutional monarchy wherein the King holds the dual role of head of state and highest religious authority, its official name is "The Kingdom of the West," indicating that Morocco situates itself culturally in relation to the place of origin of Islam, now Saudi Arabia.

Country-Specific Regulations

Morocco is a favored destination for foreign researchers. Unlike other countries in the Maghreb (i.e., Tunisia, Algeria, and Morocco, also sometimes included are Mauretania and Libya), it has a stable government, security is high, and there is an established process for obtaining permission to conduct research. Morocco is also a major tourist destination; therefore, the population is used to seeing foreigners. That said, there is also something of "research fatigue," in that some population groups are repeatedly selected for interviews and studies. All studies that rely on interviews with private persons, require permission from the Ministry of Interior. Researchers are more likely to receive the required permission if the application is submitted with the support of a recognized, registered, community-known or governmental association. Consultation with local organization leaders will help researchers identify a population(s) that is ready to participate in research, i.e., interested in the study question and not overly studied. In collaborating with local organizations, researchers will need to carefully frame study questions to reflect both scientific and local objectives. Additionally, researchers will need to incorporate local expertise in designing the protocol and describe how it will be implemented to be respectful of local customs and expectations. Several illustrations of these customs and expectations are provided below for researchers' consideration.

Addressing Ethical Issues Involving Research Participants

Moroccans are known for their generous hospitality. A researcher will frequently find herself invited for tea and cookies. This encounter is not quick but one for which Moroccans take their time. Such an invitation should not be taken as a transactional endeavor, i.e., while such an invitation may lead to contacts for interviews, one should not expect this. As trust is an essential issue when conducting personal interviews, these factors are important to consider.

The religion of Islam plays an important role in daily life in Morocco. Any researcher is advised to have a basic familiarity with Islam. Religion is a frequent topic of conversation and is not considered a private matter. Researchers need to be prepared to be asked about their religious beliefs. While there is respect for all religions of the book (i.e., Judaism, Christianity, Islam), atheism is not generally a concept that is easily understood nor accepted. The preponderance of religion in public life is often at odds with contemporary demands for secular human, minority, and individual rights. Likewise, collective cohesion is valued over and above individual aspirations (Gray, 2015).

Additionally, there is limited to no understanding of a recognized right to privacy—hence women who agree to participate in a study will believe that everything they say or do will become known to a larger public and, more importantly, to people in positions of authority. This issue has implications for an IRB; the Oral History Project can be consulted for ethically obtaining informed consent from semiliterate to illiterate population groups (Oral History Association, 2021). It will be important to gather men and women to explain the goals of any study and that the issues addressed are not pertaining to very private matters such as marital relations, i.e., sexuality. The term "wellness" is not clearly understood in a rural setting. There is a strong resistance against concepts and practices that can be perceived as being in contradiction to Islam. This perception has been the case, for example with yoga or meditation, since these practices are sometimes understood as religious practices that are not in keeping with Islam. However, as diabetes is an increasing health problem in Morocco and one that is related to eating habits, this is a topic of interest (Chetoui et al., 2018). Participants in a study need reassurance that the project is important for the health of women, children, and the entire family.

Contrary to the common belief that religion is often at odds with modernity, Chaara found that in Morocco, women's religiosity—especially in less formally educated population groups—correlates with an increased desire and interest in daughter's education and by extension overall well-being (2018). This shift in beliefs is because older women came to realize that they were held back in fulfilling their potential by an interpretation of religion that translates the religious edict of human submission to God to women's unquestionable submission and obedience to men (Chaara, 2018).

Any ethnographic research in Morocco, especially one that has women as its focus, needs to be cognizant of the patriarchal structures. Žvan-Elliott has vividly described the difficulties of a foreign female researcher, albeit one who is fluent in colloquial Moroccan Arabic (*Darija*) (2015). One of this scholar's challenges was related to the fact that she was perceived as a *bint*, literally "girl," an appellation that is used for unmarried women regardless of their age. A "girl" remains under the tutelage of her father, brother, uncle—a male guardian—and has to obtain his permission for essential acts such as going out and meeting with strangers, be they male or female. Though Žvan-Elliott was in fact married at the time of her research in a small town in southern Morocco, she faced great skepticism at claiming to have a husband who would let her stay by herself for an extended period of time in a faraway place (2015). Researchers will need to consider that family members, especially men, will have influence over a participant's responses to questions and/or ability to implement behavioral interventions. Thus, a study design that considers the family as the unit of analysis, rather than the individual, may be more appropriate.

As with the study in the Tongan community in the U.S., in Morocco too, it will be important to assure participants that the project is about health matters pertaining to the entire family in which women—the ones responsible for food preparation and eating habits—play a major role. Modern conveniences, such as pre-packaged food items, chips, cookies, soda, juice boxes, are now readily available in remote rural areas. Women who want to consider themselves "modern" may think that giving their families such snacks are a healthy choice. Likewise, there is limited understanding of the non-biodegradable aspects of trash associated with such food items. Traditionally, trash in rural areas is bio-degradable and disposed of in little heaps outside the home or along paths. As there is no organized garbage collection in certain areas, public spaces are littered with packaging materials. Therefore, it will be important to provide holistic information that includes all aspects of food preparation, purchases, and resulting waste.

Systematic reporting of any topic area is impeded by illiteracy of women. It is likely that children who attend school will do the writing. When parents are unable to read what their children write down, it is crucial that teachers are incorporated in the study and have a clear understanding of what students are asked to record, why this is important and what benefits accrue to participants. Certainly "benefits" in the poorer regions mean a tangible good, often money. Sometimes a foreign researcher is expected to provide some connection that leads to securing a job or a place in an institution of higher learning. Therefore, a researcher must be clear that she has no such thing to offer and explain straightforwardly and in practical terms what participants stand to gain from contributing to the study.

Case Study #2 (Ghana): Managing Unintended Consequences: Balancing Participant Risk and Optimal Study Design by Lisa H. Gren and Caren J. Frost

Context

In Ghana, U.S. researchers were piloting a study to determine who needed micronutrients to augment nutritional status (Gren et al., 2018). Women were the targeted participants for a comparison of two types of micronutrient supplementation linked to pregnancy and neonatal outcomes. Women received a pregnancy test and those who were pregnant were randomized to receive one of two dietary supplements. Public health nurses met with participants over a 12-month period to collect study data and dispense a monthly supply of the micronutrient. Women were thus able identify their own pregnancy status long before pregnancy symptoms would have become apparent in a regular manner. Serum samples were collected at month one of pregnancy, during the third trimester, and three months post-delivery. The researchers were surprised to see that many women who were pregnant at the beginning of the study where no longer pregnant and had not given birth by their follow-up study visits. The pregnancy loss for women in the study was 32 percent compared with roughly 16 percent pregnancy loss for Ghanaian women due to miscarriage and abortion (Ghana Statistical Service, 2009). Elective abortion is illegal in Ghana. The Guttmacher Institute reports that about 45 percent of abortions in Ghana are medically unsafe (2013).

Statement of the Issue

In this research context, information given to research participants appeared to encourage illegal activities (i.e., elective pregnancy termination) and may have led some women to seek out non-medically induced abortions that placed them at increased risk of poor health outcomes due to accessing unsafe services. Thus, there were two potential unintended consequences that the researchers had not considered that added to the legal and health risks for these women.

Outcomes

Since there was a reduction in the number of pregnant women at the beginning of the pilot study, the research team revised the study protocol. The number of women in the study who did not deliver and were no longer pregnant was twice the expected number based on national statistics. In the revised study, women were no longer screened for pregnancy. In addition, communities, not individual women, were randomized to one of two types of supplements so that all women received the dietary supplements and pregnancy was identified in a regular manner. These changes resulted in increased study costs because all women in these communities received supplementation regardless of pregnancy status. Thus, the study paid for micronutrient supplements for non-study participants as well. From a responsible conduct of research standpoint, the researchers were not putting women at risk for participating in illegal activities and/or for poor health outcomes linked to study activities.

Questions to Consider

1. How would this type of unintended consequence on a research protocol be handled in your research setting?
2. When a research team becomes aware there is an issue, what concrete steps should they take to mitigate harm to participants? For example, what entities would the team need to notify about this type of unintended consequence?

Case Study Analysis: An Indian Perspective by Hena John Fisk

Context

According to the World Health Organization (2021), micronutrients are vitamins and minerals required in very small amounts by the human body for growth and development. A deficiency of micronutrients affects health, learning ability, and productivity owing to illness and disabilities, and contributes to a vicious cycle of malnutrition, underdevelopment, and poverty (Gonmei & Toteja, 2018). India makes up one sixth of the global population out of which one-third of Indians suffer from vitamin and micronutrient deficiency (Kotecha, 2008). As Gonmei and Tojeta note, "According to the Food and Agriculture Organization report on the State of Food Security and Nutrition in the World, it is estimated that 190.7 million (14.5%) of the Indian population was undernourished during 2014–2016" (2018, p. 9). The Indian government has launched several programs or schemes to improve nutrition and health of its population. Some of these programs are Integrated Child Development Services (ICDS) scheme, Mid Meal Program, National Iron Plus Initiative (NIPI), National Iodine Deficiency Disorders Control Program (NIDDCP), and National Prophylaxis Program (Gonmei & Toteja, 2018). Even with this plethora of schemes and programs, around 0.5 percent of total deaths in 2016 were contributed to by nutritional deficiencies (India State-Level Disease Burden Initiative Collaborators, 2017).

Country-Specific Regulations

Pregnant women in India, due to their socio-economic and health status, are considered a vulnerable population for research. Sanmukhani and Tripathi (2011) define a vulnerable person as relatively or absolutely incapable of protecting their own interest. This definition includes anyone "who is very poor, illiterate patients, children, individuals with questionable capacity to give consent, prisoners, fetuses, pregnant women, terminally ill patients, students, employees, comatose patients, tribals, and elderly" (Sanmukhani & Tripathi, 2011, p. 5). In 2006, the Indian Council of Medical Research (ICMR), laid down three basic ethical principles: respect for persons, beneficence, and justice, outlining 12 general principles to guide the ethical conduct of research. These principles are voluntariness, informed consent, community agreement, non-exploitation, privacy and confidentiality, precaution and risk minimization, professional competence, accountability and transparency, institutional arrangements, principle of public domain, principle of totality of responsibility, and compliance (Sanmukhani & Tripathi, 2011). Researchers should follow the Declaration of Helsinki that states that "Medical research involving an underprivileged or vulnerable population or community is only justified if the research is responsive to the health needs and the community stands to benefit from the results of the research" (World Medical Association, 2008).

In 1980, the ICMR established Ethic Committees (ECs) throughout India. The purpose of an EC is to review biomedical research to safeguard the dignity, rights, safety, and well-being of the research participants (Thatte, 2007). In 2017, the National Guidelines for Biomedical and Health Research Involving Human Participants, expanded to include socio-behavioral research related to health and research involving biological material and datasets (Mathur & Swaminathan, 2018). To form an EC, there should be five to 15 experts, preferably with a pharmacologist, clinician, legal expert, social scientist/representative from a nongovernmental organization (NGO), philosopher, theologian, and any person from the community (Indian Council of Medical Research, 2006). Every institute where research is going on should have its own EC with its head preferably outside the institute (Sanmukhani & Tripathi, 2011). In India, there are more than 200 institutes with functional ECs for medical research. However, these committees struggle with basic issues of inadequate or no standard operating procedures, lack of trained experts, heavy workloads, inadequate space allocation, lack of administrative support, and inadequate remuneration for EC members. In addition, there is a concern about non-compliance with recommendation of the Central Drugs Standard Control Organization (Kadam & Karandikar, 2012). These issues limit the effectiveness of ECs in India.

Addressing Ethical Issues Involving Research Participants

Micronutrients have important influences on the health of pregnant women and the growing fetus. Deficiency in micronutrients can increase the risk of death from hemorrhage during delivery, lead to hematological consequences, congenital malformations, congenital abnormalities, pregnancy complications, as well as growth retardation and retarded neurobehavioral and immunological development in the fetus (Black, 2001; Census India, 2020).

In India, social hierarchies or stratifications produce maternal health inequities (Solar & Irwin, 2010). Hamal, Deileman, Brouwere and Burning (2020) identified various governmental policies on gender equality, anti-violence, abortion, and sanction of infidelity, cultural, social values, ethnicity, caste, education, income and gender as affecting maternal health. These numerous socio-economic and political factors may cause ethics-related issues in research. These factors can make it hard for researchers to truly follow the principle of voluntariness, informed consent, and community agreement and non-exploitation. It would be hard for a woman to make an independent choice without the influence of her family whether to take part in the research as the Indian family structure is based on a patriarchal system where male members of the family have greater say in any form of decision (Jat et al., 2015). Due to this, it would be hard to get true voluntary informed consent from pregnant women for their participation. In addition, the consent of the community is decided by individuals in power like the head of the panchayat/local government in rural areas. Caste, social status, education, gender, and/or religion influence participation in research studies, especially in rural settings, which can violate the principle of non-exploitation. The principle of privacy and confidentiality, too, can be compromised due to the familiarity of individuals in a rural setting.

Case Study #3 (India): Unlocking the Impact of Domestic Violence on Adult Unmarried Women in India: Challenges in Empirical Research by Aileen Savina Marques

Context

In India, with the increasing instances of cases of domestic violence, civil society organizations expressed the need to identify and unlock the impact of domestic violence on adult unmarried women. In the traditional patriarchal Indian household, young adult unmarried women seldom have opportunities to express their views. Decisions related to their welfare, marriage, education, dressing style, vacation, etc. are generally made by the male elders of the family without any consideration to the choice and consent of young adult unmarried women. Young unmarried women are compelled to abide by the dictates of the older men in the family as older women are not considered to hold any position of authority within the family. There exists a thin line between discipline and violence. The curbs on individual autonomy and personal liberty are most often seen as discipline rather than violence.

Research to understand domestic violence on young unmarried women will require addressing ethical concerns. In 2011, the overall literacy rate in India was 73 percent, with the female literacy rate being 64.6 percent (CensusInfo India 2011-Final Population Totals, n.d.). While literacy and level of education form the basic indictors of the level of development achieved by the society, the low female literacy rate is an indicator that women are not encouraged to pursue education in spite of social welfare schemes like *beti bachao, beti padao* ("protect the girl, educate the girl") introduced by the central government. Illiteracy and lack of opportunities mar a woman's ability to comprehend the social, political and cultural environment in which she lives. Informed consent for any research by women, specifically young unmarried women is one of the major challenges a researcher will face.

Participation in a study might mean spending more time outside the house and, in some cases, interacting with strangers, thus making it rather difficult to obtain permission to stay out longer. Denial of the right to personal liberty of young unmarried women, though unconstitutional, continues unchallenged in most Indian households. The fundamental right to freedom of speech and expression is enshrined in the Indian Constitution. However, the social and cultural context with specific reference to young unmarried Indian women does not encourage freedom of speech and expression. Women who choose to exercise their right to freedom of speech and expression are often labelled *dabbang* (bold or assertive). The dynamics of caste and class in India continue to dominate the social life. It is pertinent to note that 41 percent of women belonging to the scheduled castes and tribes (SC/ST) faced physical violence, which is much higher than for women who do not belong to these categories (Roa, 2015).

Statement of the Issue

A hypothetical study was aimed at identifying and assessing the impact of domestic violence on Indian adult unmarried women aged between 18 and 25 years. The Protection of Women from Domestic Violence Act (PWDVA) of 2005 in India provides for the effective protection of the rights of women guaranteed under the Indian Constitution who are victims of violence of any kind occurring within the family. Women whether married or unmarried who are part of the family who are victims of any form of abuse (physical, emotional, verbal, economic, and/or sexual) are considered aggrieved persons as defined in section 2(a) of the PWDVA. In this research context, information given to research participants may seem to be considered as a stimulant for young unmarried women to respond or react to domestic violence, which they unconsciously are conditioned to accept from the subtle statements that girls "don't talk/laugh loudly" and "marry the boy we choose" to "men can control women with violence."

The proposed study on the topic of domestic violence-initiated conversations to identify and respond to domestic violence, which may have led to an increase in reporting of instances of domestic violence by young unmarried women. The possibility of early marriage of these young unmarried women to "silence" them or "tame" them cannot be ignored. Young unmarried women may question the patriarchal power structures in their families, making them more vulnerable to domestic violence.

Outcomes

The subtle and rather invisible dimension of domestic violence faced by young unmarried women (between 18 and 25 years) has come under scrutiny and might be linked to the hypothetical study noted earlier. Young women may quantify the impact of domestic violence and begin conversations with the elder women in the families about domestic violence. Study participation may initially create some unrest within families—and may extend out to communities—but may also enable younger women to start conversations about their rights in society.

Questions to Consider

1. How will "parental supervision" of young unmarried women in India be handled by the researcher considering that these women are to be "under the care and custody of their fathers/male elders"?
2. Will it be practically viable for a researcher to adhere to ethical research guidelines, such as obtaining informed consent, during the process of data collection from young unmarried women who are reluctant to assert personal liberty?
3. Would unmarried young women actually self-identify as victims of domestic violence in this research project? How would you ascertain if the information shared as part of the research by young unmarried women is truthful and correct considering that the Indian socio-cultural context does not allow one to bring the "private" information into the "public domain" for public scrutiny?
4. How would a researcher initiate a rapport-building process to inculcate trust in the participants so that the information provided by the participants will not lead to disgrace or affect family's *izzat* (dignity)?
5. Would you tolerate a potential increase in the short term in domestic violence as a consequence of your research?

Case Study Analysis: A Perspective from Pakistan by Ronak Iqbal

Context

The British ruled the Indian subcontinent for 89 years from 1858 to 1947. The separation of the states into two nations was based on religious differences between Hindus and Muslims. So, any researcher must understand the impact religion has on Pakistani society. In addition to the issues discussed in the case study from India, one should consider the following additional viewpoints from conservative Muslim-majority Pakistani society.

Country-Specific Regulations

While the Pakistani government has a National Bioethics Committee (NBC), constituted in 2004, there is evidence that there is no consistent oversight or standards that are adhered to (Jafarey et al., 2012; Sherin, 2013). The relevant principles that should guide the development of national standards have been identified by Pakistani researchers as the Declaration of Helsinki (World Medical Association, 2013), the Belmont Report (DHEW, 1979), and the guidelines published by the Council for International Organizations of Medical Sciences (CIOMS, 2016). In the absence of national standards, research ethics committees (RECs) have, in some cases, developed their own guidelines to assist researchers in going through an ethics review by explaining how the committee works and its guiding principles, and providing templates for researchers to complete in preparation for a review. One useful example is provided by the Institute of Business Administration Karachi, situated within the School of Economics and Social Sciences (Ejaz, et al., n.d.).

Recently, in recognition of the collaboration between researchers in high-income countries and low- or middle-income countries (LMICs), there have been efforts to align ethical principles among the researchers. The Global Code of Conduct for Research in Resource-Poor Settings (GCC) was created with the goal to promote equitable research partnerships (TRUST, 2018). The GCC is built on four principles: fairness, respect, care, and honesty. Each principle is associated with articles—a total of 23—that can guide the development of an equitable research relationship for entities that may start off at different levels of power. Chatfield et al. (2022) evaluated collaborations of researchers in the United Kingdom with researchers in Pakistan and in India. They found that there were challenges for implementation, but that application of the GCC promoted equity in these international research collaborations.

These same four principles offer a useful framework for thinking about ways to balance power among young unmarried women, the men who have decisional capacity over so many facets of their lives, and the researchers wanting to study the health implications of domestic violence. For example, the GCC principle of fairness promotes relevance and potential benefits for the community in which the research is taking place. The GCC principle of respect promotes the engagement of local co-researchers (both academic and cultural) to ensure that implementation aligns with local customs and understandings. Engaging all stakeholders in a process that properly situates research on domestic violence should both identify relevance and potential benefits, as well as suggest ways to implement the research in a culturally acceptable manner. The GCC principle of care promotes valuing the individuals who participate, including adding safeguards for physical and mental safety, as well as protecting the information they provide for study purposes. As part of this principle, the GCC suggests implementing a mechanism for participants and community members to voice complaints. Attending to how the research is affecting people, while it unfolds, could help to mitigate the potential for increased domestic violence in the short term, as a result of women's participation.

Addressing Ethical Issues Involving Research Participants

One specific example in the Pakistani context that could benefit from applying the GCC principals is that the definition of domestic violence against unmarried women needs to be better outlined. Do Pakistani women understand and define domestic violence the same as the researcher? Being pushed around, shaken to the core, having things thrown at them, being slapped or punched, having limbs twisted, hair pulled, and/or being dragged across the floor may just be considered a part of gender disparity and may not be considered domestic violence. A researcher must consider these potentially different views and include local stakeholders when developing the data collection tools to learn about participants' experiences.

Another consideration where application of the GCC principles could be relevant is the practice of honor killing, also known as *Karo Kari* (Amnesty International, 1999; Anam, 2019; Patel & Gadit, 2008). In this patriarchal culture, women's lives are structured through strict maintenance of the honor code. The researcher must frame questions that acknowledge the practice of *Kori Kari* for some communities. Honor is a culturally complex concept and a researcher would need to understand how a person and family might lose their honor. In addition to *Kori Kari*, other acts and threats of domestic violence, such as the threat of being burned alive or having acid thrown at the face for disfigurement, has been observed in some communities where honor appears to have been threatened or lost. When enrolling a participant who has scars from past abuse, a researcher must strive to overcome the cultural barrier when asking such difficult and intimidating questions. Due to the study participant's history of abuse, she may be reluctant or unable to answer. One solution researchers have used is to include family members in generalized discussion about domestic violence. This approach may seem to provide anonymity to victims (who are the participants in a study) and allow them to answer freely without being identified, but at the same time it may increase the risk to the woman, other family members, or the researcher depending on who participates in the discussion—as part of the principle of care, researchers will need to anticipate potential harmful situations and have plans in place to protect these stakeholders.

More broadly, cultural traditions and messaging may operate at the community level and not just the family level. For example, in Pakistan religion plays a vital role in what are acceptable behaviors. The practice of marrying a girl child to the Quran still exists in some rural and conservative communities, which can deprive girls of their property rights. It ensures that the girl child will not bear children who demand a share in the family property (Amnesty International, 1999). The age of marriage is also a question that needs to be considered. In many subcultures of Pakistani society, girls are married off as soon as they get their first menstrual cycle. Due to the proximity to the equator, girls hit menarche earlier than their Western counterparts. Some of these girls can be as young as eight years old (Nasrullah et al., 2014).

A second example is the important role of media and television in Pakistani society. We must consider the implication of women's representation in the standard television series and movies produced in both Bollywood and Lollywood (Kamran,

2019). Some of the most prominent concepts around honor, including economic roles, marriage, domestic violence, silencing of victims of sexual harassment, rape, domestic violence, etc. come from portrayal in the media (Mentzell Ryder, 2015; Zubair & Zubair, 2017).

Questions to Consider

1. If the research definition of domestic violence makes the study participant aware of their status as a victim of abuse, how will the researcher manage potential mental health consequences for an individual who now has to accept this knowledge?
2. What can a researcher do to reduce or mitigate risk for women who participate in such a research study as noted above?
3. Is being married off as a young girl to the Quran considered an act of domestic violence? How will the researcher deal with this social phenomenon?

Summary of Chapter

It has long been recognized that the participation of women in medical research has been limited both from the research deployment and participation perspectives (Mastroianni et al., 1994). While not a vulnerable population per se, improving and ensuring the participation of women in research requires considerations that should not be deemed "special"; rather these considerations should become routine to the daily discourse of research activities. In this chapter, we have examined the traditional and interrelated roles of women in culture and how failing to understand these roles can endanger the health of women and family members, and jeopardize the validity of research.

In the U.S., we saw how the participation of women in the Polynesian community resulted not only in intended changes in the woman's behavior, but also unintended changes in the environment for family members dependent on the women's performance of traditional and cultural roles. Fortunately, these unintended consequences likely ended with a healthier household and family in the case discussed above where exercise and dietary habits were improved for the entire family. This case demonstrates the potential ripple impact on other family members related to the role of women in setting the circumstance for health behaviors at home, and should be a consideration for research design and analysis.

In Ghana, we saw how research focused on the health of women prior to and during pregnancy was jeopardized when early pregnancy identification collected for the study prompted some women to risk their lives and future fertility by seeking unsafe practices to terminate their pregnancies. The IRB-approved protocol had not considered actions taken by research participants that endangered them as a result of information they received during the course of their research participation. Historically, the recommended action by the IRB in cases where the potential for harm was identified was to continue with the study unchanged until the protocol modification was reviewed and approved by the IRB. The apparent contradiction between safeguarding the health of research participants through adherence to study protocols and the need for the freedom to respond immediately and appropriately to observed dangers has since been incorporated into the protocols of the IRB. However, unforeseen behaviors as a result of participation in research is likely to continue. A need for constant vigilance, reporting, response, and remediation of the dangers of participation in research is requisite on both the researcher and the governing body overseeing the ethical conduct of research, and it is necessary that the relationship continue throughout the course of a research activity in order to assess whether, and how, risk to participants is changing.

Finally, the case study from India forces us to address the complex cultural definitions for domestic violence. We see how cultural norms can alter what behaviors are defined as domestic violence and lead to discrepancies in whether women identify themselves as having experienced domestic violence. If, during the course of the research, a woman comes to identify as a victim of domestic violence, researchers need to consider how this may impact her future wellbeing. Her identification as a victim—with its attendant influence on her mental and emotional health—also means that someone else is identified as an abuser. The person labeled as an abuser may actually escalate the physical abuse, as punishment for having participated in a study that brought these events to light. Similarly, the woman may have limits placed on her independence, which can also negatively impact her well-being. Perhaps the most troubling and difficult questions raised by such research include: Does participating in research on domestic violence imply victimhood? Does participation then risk an increase in the frequency and severity of future abuse? To further our understanding of, and the extent of abuse, are we willing to accept an increase in abuse? If so, for how long, how severe, and for what research purposes are we willing to accept an increase?

Taking the time and effort to understand a woman's role in her culture, traditions, and broader society is critical to improving women's participation in relevant and safe research. Researchers should actively seek to include women in all aspects of research development and interpretation. Researchers will find it necessary to consider the complexities of the cultures and traditions that women operate in, exert influence over, and are subject to when designing, conducting, and participating in research activities.

Chapter Exercises

1. When considering behaviors, cultural context is a vital consideration to adequately conduct, participate in, and understand research. What impact does culture play on gender roles, expectations, and the participation of women in research?
2. Establishing trust with a participant is vital to obtain accurate information in research. This trust can take an extended period of time to build and establish. How does one build trust in a community different from their own? How long must one work to establish this trust? Does including a member of the local community in the research project facilitate this trust or increase the danger of a breach of confidentiality?

3. The purpose of research is to provide additional information to less well-understood topics. At times, the investigation into these areas may uncover additional or unintended issues that pose a risk to the researcher or participants in the project. What obligation does a researcher have to anticipate unintended findings? How does one respond to unintended findings when partnered with community participants who may have an interest in the research and its outcomes?
4. Protection of participants, understanding and improving people's lives, health, and safety is often the motivation for research. However, at times, the topic of a research project may focus on aspects that are dangerous to women participants. Domestic violence is one such example. How prepared should a researcher be to accept the potential for a small increase in the short term of violence from participation in a study, if the overall project seeks to understand, diminish, or end domestic violence? If so, what is an acceptable small increase, and how long is short term?

4 Male Participants

White, heterosexual, middle-class males between the ages of 35 and 55 years have typically been the primary subjects in research studies (Dresser, 1992; Konkel, 2015). As an example, drug clinical trials skew toward white participants—some as high as 80–90 percent—even though 40 percent of the U.S. population identifies as a racial or ethnic minority (The Editors-Scientific American, 2022). While the median enrollment of men and women in U.S. clinical trials was roughly equal between 2000 and 2020, disparities remain within specific disease categories in terms of the proportion affected and the proportion enrolled in research studies (Steinberg et al., 2021; van Diemen et al., 2021). Steinberg et al. (2021) identified the greatest under-enrollment disparities for men occurred in clinical trials of musculoskeletal conditions and trauma, and head- and neck-related conditions (otolaryngology); areas of lesser disparity included clinical trials of mental health, pulmonary, congenital, and nutritional/metabolic conditions.

The United Nations Sustainable Development Goals advocate for good health and well-being for all (goal 3) and gender equality (goal 5), recognizing that gains in one goal support gains in other goals (United Nations, n.d.). In a 2019 editorial in *The Lancet*, it was noted that life expectancy for men is 4.5 years lower than it is for women. The editorial noted that addressing the disparity in life expectancy would require addressing health-care access barriers and health-related behaviors that are influenced by gender (Lancet, 2019).

Two examples—HIV/AIDS and mental health—are illustrative of the need for research to address health-care access and behavior in improving men's health. In the U.S., men far outnumber women in enrollment in clinical trials for HIV/AIDS (Smeaton et al., 2020), and the knowledge generated from these types of trials informs care delivery. Yet, there are a number of areas (i.e., sub-Saharan Africa, the Middle East, and Russia) where societal norms and beliefs surrounding men who have sex with men (MSM) and HIV/AIDs have hindered prevention and treatment efforts for these groups (Mills et al., 2012). Research on behavior and social stigma are needed to improve access to care and improve health for men at risk of, or diagnosed with, HIV/AIDS in these settings. Regarding mental illness, the overall prevalence is similar between men and women; however, types of disorders differ by gender, with men having higher rates of substance use and antisocial personality disorder and women having higher rates of eating disorders and anxiety (Resnik & Jones, 2006).

DOI: 10.4324/9781003033233-4

These differences in prevalence, combined with the under-enrollment of men in mental health research (Steinberg et al., 2021), means that our understanding of risk factors and knowledge of successful interventions could be improved by having studies adequately powered to study the impact of gender.

Where researchers identify under-representation by gender, van Diemen et al. (2021) recommend consideration of motivators, facilitators, and barriers related to study participation. They recommend (1) creating study teams that reflect diversity; (2) working collaboratively with diverse communities to understand their experiences on the topic of interest; and (3) using the information gained to develop gender-responsive educational, advertising, and informed consent materials, as well as providing ongoing training to research and clinical teams on diversity issues in health research and clinical care.

The case studies in this chapter consider the following topics:

- how policy drives the research agenda and can contribute to prolonging the classification of a group as an underserved population;
- instituting standardized laboratory procedures to minimize disease misclassification in multi-site research settings;
- the impact of stigma, and associated social consequences, on clinical trial operations.

Case Studies and Analyses

The case studies in this chapter are based in three countries: U.S., India, and Malaysia. The case studies deal with the interplay between broad policy or cultural expectations and how researchers adapt protocols to ensure that research is conducted in a rigorous and ethical manner. The first case study describes research restrictions put in place as a result of a federal policy that banned transgender men in the military from disclosing this information at the risk of being discharged (U.S.); the second describes the challenge of conducting multi-site research when laboratories are not required to conform to a federal quality standard (India); and the third describes approaches for addressing privacy and confidentiality related to enrolling and providing incentives to men who have sex with men (MSM) in a setting where this group is highly stigmatized (Malaysia). Through these case studies, we see how politics and culture shape the context within which researchers implement studies. Researcher adaptations included (a) pausing research on how military service is protective for the mental health of transgender men until policies that harm these men's careers were reversed, (b) considering the balance between making research more broadly available geographically against the need for ensuring that laboratory procedures meet quality standards that could be difficult with existing infrastructure, and (c) changing enrollment procedures to use trusted clinical contacts as the people to introduce the study rather than the researchers themselves.

The case study analyses are provided by researchers in two countries: South Africa and the U.S. The third analysis is left for readers to formulate, drawing upon their review of the Declaration of Helsinki and the Belmont Report, as well as the questions that accompany each case study. The case study analysis from South Africa points out that even when government policy is supportive of research with a sex- and gender-minority group, such as transgender men, implementation can still be difficult. The researchers point out the importance of maintaining participant privacy and confidentiality in order to prevent negative consequences for participants. These negative consequences could impact career advancement, as in the case study, and social status. By being reliable partners in protecting participant privacy and confidentiality, researchers strengthen their standing as capable implementers. The case study from the U.S. identifies standards from one country (in this case, the U.S.) that could be adopted by researchers in other countries, such as India, where there are not widely accepted certification standards for laboratories. The ability to implement quality standards in outlying areas—in this case, Tier 2 and 3 cities—positions these locations to continue to be involved in multi-site studies. Such upgrades in infrastructure will allow people in additional areas to be involved in research, making the potential benefits more broadly available.

Case Study #1 (United States): Don't Ask, Don't Tell, and Don't Conduct Research that Could Save Men's Lives by Charles P. Hoy-Ellis

Context

Transgender individuals' gender identity is at odds with their biological sex. Sex—designated as male or female—refers to the biological aspect of human bodies, such as reproductive organs and chromosomes. We assign sex at birth based on these characteristics, which becomes legally binding throughout life. Designation of sex is required on everything from birth certificates to drivers' licenses to passports and can only be changed by the courts. Gender is psychological and cultural in nature and is traditionally predicated on one's assigned biological sex at birth. U.S. culture overwhelmingly divides gender into an assumed "natural" binary of boys/men (biologically male) and girls/women (biologically female), despite overwhelming research in the natural world that sex and gender are fluid (Roughgarden, 2013). Culturally, American conceptualizations of gender have both expectations and proscriptions—boys/men should and should not do this or that, girls/women can do this but can't do that.

Psychologically, gender is our deepest sense of ourselves as girls/women or boys/men, *regardless* of our biological sex of female or male. Transgender men, assigned a female sex at birth, nonetheless experience themselves as men. Nearly half of transgender people in the U.S. experience significant depression and suicidality, due to social stigma and psychological–biological dissonance (Hoy-Ellis & Fredriksen-Goldsen, 2017). Many American transgender men utilize gender-affirming medical care (e.g., use of testosterone, mastectomy) to more closely align their bodies with their gender identity, which has been shown to improve mental health and social functioning (Goldstein et al., 2020). Initially launched in 2010, a large, nationwide American community-based research project included a significant subsample of transgender adults aged 50 years and older. Results provided preliminary evidence that older transgender men who had served previously in the U.S. military had a lower risk for depression compared to those who had not, and that gender-identity factors played a mediating role in this relationship (Hoy-Ellis et al., 2017).

Statement of the Issue

In the U.S., transgender individuals serve in the military (i.e., Army, Navy, Air Force, Marines & Coast Guard) at high rates. Over a five-year period, transgender individuals' status regarding their service changed multiple times (Baldor, 2021). In 2016, the U.S. Department of Defense issued a ruling that would have allowed transgender Americans to serve openly in the military and to receive transition-related medical care. In 2017, the Trump administration rescinded that ruling. Transgender men could no longer serve openly, and transgender men who had received a diagnosis of gender dysphoria, which is a prerequisite to obtaining transition-related care, could no longer enlist in the military. This policy was a significant barrier to further research in understanding how military service can be a protective factor in transgender men's mental health. Notably, the transgender ban was reversed by the Biden administration in 2021.

Outcomes

The Trump-era policy banned transgender men from serving openly in the military, remaining in the military if they received a diagnosis of gender dysphoria while there, or enlisting if they had a diagnosis of gender dysphoria. Transgender men are among the most vulnerable and stigmatized populations in the U.S. Designing and implementing a prospective longitudinal study to better understand the role of military service in transgender American men's mental health would put potential participants at unacceptable risk during the period the ban was in place. At the same time, not developing such a protocol presented a frustrating delay in research that could inform the development of prevention and intervention efforts that could improve the mental health and well-being of these men.

Questions to Consider

1. What ideas for the development and implementation of a research protocol to address this issue might your research team identify in your research setting?
2. What specific ethical issues and principles would your research team identify in this dilemma? What ethical obligations would your team need to follow?

Case Study Analysis #1: A Perspective from South Africa by Busisiwe Purity Ncama, Zamasomi Prudence Busisi Luvuno, and Sinegugu Evidence Duma

Context

In the continent of Africa, South Africa is the only country that recognizes the rights of transgender people. These rights are protected by Section 9 of the

Constitution of the South African Republic (1996) and the Promotion of Equality and Prevention of Unfair Discrimination Act 4 of 2000. The latter was enacted to give effect to section 9 of the Constitution—to prevent unfair discrimination and harassment of people based on their sex, gender, or sexual orientation. The Act further defines "sex" to include intersex. South African transgender people can apply to change their sex description in the birth register in terms of the Alteration of Sex Description and Sex Status Act 49 of 2003. The Act states that, "Any person whose sexual characteristics have been altered by surgical or medical treatment or by evolvement through natural development resulting in gender reassignment, or any person who is intersexed may apply to the Director-General of the National Department of Home Affairs for the alteration of the sex description on his or her birth register."

Transgender people in the South African military or the South African National Defence Force are protected by the Defence Act 42 of 2002, Chapter 17, Section 105, which states that any member of the Defence Force or the employee who discriminates against a person based on their sexual orientation is guilty of an offense and could, if found guilty, be imprisoned to the maximum of five years. This Act is significant progress considering that during the apartheid era an aversion project existed in South Africa—a medical torture program that targeted homosexual soldiers and identified them as mentally ill, conscripted them into the program in the South African Defence Force to "correct" their sexuality through castrations and electric shock (Kaplan, 2001).

Similarly, in the U.S., a 2016 ruling that allowed transgender Americans to serve openly in the military and to receive transition-related medical care issued by the U.S. Department of Defense was rescinded by the Trump administration in 2017. President Biden's Executive Order of January 2021 lifted the ban (Biden, 2021). This ruling by the Biden administration renders the American situation comparable to South Africa, thus similar research on transgender men in the military can be conducted. In reality, South Africa often has good policies and legal frameworks on paper but the implementation is usually a challenge; it would, therefore, be prudent to conduct research in the area of implementation of these policies.

Country-Specific Regulations

In South Africa, most research conducted among transgender men has been on human rights and access to health services. The research is usually conducted using surveys, probably to protect the identity of the participants. Very little, if any, research has been conducted on protective factors for development of gender dysphoria, minority stress, and other related mental health issues affecting transgender men. Future research should focus on longitudal and prospective studies on transgender men in different settings, including the military and the universities. Qualitative and participatory action research designs which allow transgender men to own the process of knowledge construction about their needs should be considered by research teams in South Africa.

Addressing Ethical Issues Involving Research Participants

Although protected by the South African Constitution, transgender men remain the most vulnerable and stigmatized population. Transgender men often experience internalized social stigma and constant expectation of rejection which is socially constructed and shaped by cultural and religious beliefs (Hatzenbuehler & Pachankis, 2016; Meyer, 2003). The protection of transgender men should remain a priority throughout the research process. The research team wanting to conduct research on transgender men should be guided by the general principles of the Declaration of Helsinki (World Medical Association, 2013) as a framework that ensures that their research remains scientifically and ethically sound. One of the benefits for using the Declaration of Helsinki framework is that it promotes participants' safety irrespective of the type or length of the research design. The key principles are described further below.

Autonomy: The Right to Self-Determination and the Right to Respect

Researchers interested in conducting research on transgender men often face challenges in recruiting transgender persons using normal recruitment procedures. The transgender men are not afforded the right to self-identification/determination when completing employment forms or healthcare admission forms. Instead, they are forced to use male or female binary definitions of gender. Transgender individuals are not easily identifiable within the military, since it would be based on self-declaration and the application forms do not explicitly ask for gender or sexual orientation, but only biological sex. Stigma also prevents disclosure by transgender people in an environment with people who promote a macho persona. In this case, the researcher may have to adopt different recruitment strategies, including adjusted snowballing, using social media requesting transgender men to anonymously make contact with the researcher, and accessing them through community-based support groups for transgender men. These recruitment strategies may ensure that the transgender men's autonomy is respected during recruitment and throughout the research process.

Confidentiality: Ensuring Security, Privacy and Protection of Patient's Information

Transgender persons often experience gender dysphoria and other common mental health disorders associated with stigma. The principle of confidentiality may prevent the medical practitioner from sharing the diagnosis, or any health-related information with any of the research team members wishing to access information about the transgender person's medical history and care used for transitioning. Strategies for assuring confidentiality in facilitating the transfer of medical data to researchers should include clearly delineating what information will be collected and how it will be accessed. Topics for discussion might be: (1) defining what information is sought and justifying why its collection is necessary to the research; (2) setting up user identities, roles, and log-ins, as well as appropriate points of access; (3) patient consent or reconsent processes; and (4) data storage security measures and timelines for destruction of stored data.

The research agenda on transgender men in different countries is shaped by the politics of the day. It has emerged that each administration that comes into office responds to the plight of transgender men differently. Researchers should consider the political situation in knowledge production. Research into protective factors for development of gender dysphoria, minority stress, and other related mental health issues affecting transgender men is scarce. This situation means that the care of transgender men is not evidence-based since they are excluded from research. This exclusion violates the principles covered in the Declaration of Helsinki in relation to transgender men.

Case Study #2 (India): Deviations to Study Criteria of Serological Laboratory Investigations of Target Populations: Effects on Participant Safety and Evaluation of Study Outcomes by Preethi Shivayogi

Context

Quantitative measurement of infectious viral load assays at clinical diagnostic laboratories and their interpretations can vary across cities on the Indian subcontinent with ensuing consequences to safety and study outcomes. With the expansion of research to new locations, smaller cities have become involved in research (i.e., Tier 2: 50,000–100,000 population, Tier 3: 20,000–50,000 population). It has been established that diagnostic laboratory processes and services are not uniform or of optimal quality in Tier 2 and 3 cities, with differing bioassay use and limits of detection. Comparison of virological laboratory assessments across multiple locations requires precision process techniques with calibration. However, laboratory diagnostic services are not currently subject to routine inspection and quality control.

The incidence of viral infections is generally high in Tier 2 and 3 locations, including for men with HIV infection. Recent estimates suggest that HIV incidence is increasing among MSM (Patel et al., 2021). Seroprevalence with quantitative measurement of viral loads plays a significant role in measuring the health status of study participants and allowing comparison across study sites. Discrepancies in laboratory sensitivity and specificity of viral loads leads to protocol deviations and places investigators in a quandary regarding interpretation of these viral titers for making treatment and safety decisions, as well as linking viral titers to other research endpoints.

Statement of the Issue

There are several scenarios faced by site personnel in relation to study-related laboratory inconsistencies: decisions regarding participant eligibility, what treatment is appropriate based on viral load, and confidence in reporting study findings.

Outcomes

In India, declines in HIV incidence have been noted over time among heterosexual individuals but not among MSM (Patel et al., 2021; World Bank, 2012). Changes in prevalence of detectable viraemia (PDV) are associated with HIV incidence at the community and individual level (Patel et al., 2021). Consider a study to assess different methods for identifying and recruiting HIV positive men in community settings and enrolling them in Treatment as Prevention (TasP), as a way to reduce transmission of HIV (USDHHS, n.d.). Baseline viral load would be used to determine participant eligibility, and longitudinal assays would be used to assess compliance with treatment.

Potential participants with false negative laboratory results (i.e., viral loads) would not be enrolled and provided timely medical intervention if they were classified as not being infected with HIV. Conversely, enrolled participants with false positive laboratory results (i.e., viral loads) would be exposed to investigational drugs and their potential side effects. Further, the potential benefits of treatment might not be accurately identified if a large proportion of study participants was misclassified with regard to their HIV status, confounding the ability to assess the benefits and harms of TasP, or of any one particular medication. Participants with equivocal viral loads may require repeat testing, adding to participant burden (i.e., time, transportation, and any associated procedural risks), and overall study cost. Each of these unintended consequences can be associated with increased health risks for participants. Further, research safety and outcomes data could be erroneous, leading to false reports to study sponsors and potentially to inaccurate conclusions about study hypotheses. Therefore, stakeholders involved in research in Tier 2 and 3 cities should consider the availability of accredited clinical laboratories that use standardized protocols and materials, and diligently apply selection criteria in determining which laboratories are involved in research protocols.

Questions to Consider

1. How should deviations and discrepancies in laboratory tests be handled in research settings by stakeholders?
2. When research teams become aware of issues related to reliability of laboratory findings, what concrete steps should they take to mitigate harm to participants and effects on the evaluation of safety?
3. How should interpretation of study endpoints be explained by study teams when the possibility of laboratory error exists?
4. How do your answers to these questions help you develop criteria for judging which laboratories should be included in a multi-site research protocol?

Case Study Analysis: A Perspective from United States by Margaret Carlson

Context

One of the benefits of clinical trial participation is access to treatment that might not otherwise be available. In India, opening up studies in Tier 2 and Tier 3 cities is one way to expand access to clinical trials. However, the case study highlights the challenge of working with labs that do not meet standardized criteria, leaving certification of laboratory quality in the hands of researchers. In the study, viral load assays were proposed as a way to determine who was eligible for study participation, and monitor viral load over time to answer questions about safety and efficacy of TasP to reduce community-level HIV transmission.

In the U.S., eligibility for clinical research studies is carefully assessed to ensure that those who do not meet criteria are not subjected to unnecessary medical treatments. Eligibility may be based on age, gender, and symptoms presented. Other eligibility criteria may be determined from laboratory results such as tests of urine or blood, or in this case a viral load assay. Rigorous adherence to eligibility criteria is necessary to protect participant safety and ensure that enrollment supports the research goals. Each machine used for laboratory testing has a specificity and sensitivity score and these should be considered when designating equipment that is acceptable for use in research studies. These numbers may limit the range of those who can be enrolled.

Regulatory protocols governing biospecimen collection and analysis should be standardized across all clinical sites. For study sites with a small number of participants, it may be possible to have the same person collect and perform all assays, while larger sites will have to rely on standardized protocols and operating procedures to reduce variability between analysts. Such standardization is critically important when the study involves multiple sites, as this case study does, so that each site follows the same protocol (i.e., the Standard Operating Procedure or SOP), although some procedures may be individualized by site. For example, the *protocol* may specify that viral load be assessed and then the materials and samples be discarded according to local procedures for handling biohazardous material. While this protocol would be followed by all sites, each site might would have its own operating *procedures* for site-specific operations about how assay materials are stored, procedures for protecting study personnel who work with biohazardous material, and how biohazardous materials are disposed of. Thus, sites would enumerate both the study protocol and any site-specific information about procedures for following the protocol into that site's Manual of Procedures (MoP) or Manual of Operations (MoO). The MoP/MoO should then be approved by the appropriate regulatory body(ies) before participants are enrolled, both to ensure the safety of participants and study personnel, and to specify standards for sample collection and analysis that result in high-quality data.

Country-Specific Regulations

In the U.S., an Institutional Review Board (IRB) is responsible for reviewing research involving human participants—including review of protocols (i.e., SOPs)—to ensure these documents comply with current laws and ethical standards. One such ethical standard in the U.S. is that all labs that produce data used in clinical decision-making need to be Clinical Laboratory Improvement Amendments (CLIA) approved, thus ensuring the quality of testing (CMS, 2022). This information is documented with current CLIA certification, granted by the U.S. Centers for Medicare and Medicaid.

Other federal regulations stipulate rules for institutions that conduct research, in order to ensure that personnel are adequately trained to conduct human subjects research. Institutions must have a federal wide assurance (FWA) number that indicates they comply with ethical standards for human studies. Personnel are certified on Good Clinical Practice and Human Research Curriculum to ensure they are prepared to address informed consent, data collection and storage (including Health Insurance Portability and Accountability Act (HIPAA) regulations specific to data collected in the course of clinical practice), participant privacy and confidentiality, working with vulnerable populations (such as children or those with diminished capacity for consent), etc.

Addressing Ethical Issues Involving Research Participants

In this case study, the use of an SOP accompanied by training of personnel at each site is one way to standardize the data collection in Tier 2 and 3 cities, even in the absence of a national accrediting body for the laboratory. Routine observation of study procedures can be used to ensure adherence to the protocol. Random selection of biospecimens from each site for quality assurance reassays can be used to check the accuracy of the test results. However, researchers need to consider approaches to minimize degradation of samples over time or due to transport conditions.

When consenting an individual into a research study, the procedures involved should be discussed in detail, including the associated risks and benefits. The participant must be made fully aware of the possibilities—both beneficial and harmful—before the informed consent process is considered completed. Participants should be notified whether they will receive laboratory results, and potential outcomes associated with laboratory results should be part of this discussion, with assurances that participants will be notified in a timely manner. Two potential issues related to laboratory results are described and discussed below relative to the hypothetical study of TasP proposed in the case study.

A false positive at baseline would result in enrollment of a man who was not truly HIV positive. If this were discovered at a quality check following enrollment, the participant should be notified, thanked for his participation to that point, and treatment should stop. One way to potentially mitigate inappropriate enrollment at study sites with laboratories that are new to involvement in clinical trials would be to have a two-stage consent process, where a screening viral assay is completed at the new lab, with a second assay conducted prior to formal enrollment at a lab known to be in compliance with the requisite quality standards. This process will add time and expense to the conduct of the study but may reduce the enrollment of men with false positive viral loads.

A second issue that could arise in this study is erroneous lab results—whether falsely low or high—which may alter clinical decisions. If this error is systematic (e.g., always high, or always low), then it may bias the results in one direction or the other and lead to erroneous conclusions. If this error is at random, then it will dilute the impact of the treatment effect, meaning that a larger sample size will be required to reach conclusive results about the hypothesis. Such a situation results in more expensive studies, longer time to results, and a greater number of participants exposed to potential harmful effects.

Once discovered, errors such as those described above (i.e., protocol deviations) should be rectified. For individual participants, notification should occur and assistance provided to address any health issues that arise. A note to the participant file should document the deviation, the conversation with the participant, and appropriate steps for resolution. In multi-site trials, this would additionally require reporting to the central data steward. At the site level, a root cause analysis should occur, and changes to procedures documented in the MoP/MoO or SOP that will be used in the future to prevent the deviation from continuing to occur; these updates should be accompanied by staff training on the MoP/MoO/SOP changes.

As researchers gain experience, they will likely anticipate these types of situations in advance and develop the initial protocol to address them. However, all researchers should be prepared to quickly address protocol deviations as they arise.

Case Study #3 (Malaysia): The Intricacies of Working with HIV Stigma and Vulnerable Populations in a Multi-ethnic Country by Ng Chirk Jenn and Nishakanthi Gopalan

Context

In Malaysia, a group of researchers from University A decided to conduct a qualitative study to explore the health-care experiences among patients living with human immunodeficiency virus (HIV) attending a government health clinic. These patients were mainly young men who have sex with men (MSM) and the majority were students from University B. The study was motivated by the high default rate (failure to complete recommended health-care visits) observed among this high-risk population, which affects the prognosis of the illness. The research team obtained permission from both University A and the Institutional Review Board of the Ministry of Health before beginning the study. The doctor in charge of a clinic welcomed the team to conduct the study in the clinic. She instructed her staff to give the research team full cooperation. However, the research team struggled to recruit any participants. They decided to change the recruitment strategy and request the help of a medical assistant at the clinic who oversees HIV services in the clinic. The recruitment improved significantly after the medical assistant began helping to recruit participants and obtaining informed consent for the study. Despite agreeing to participate, one of the participants refused to put down his name on the consent form, as he was concerned about people learning of his HIV and MSM status. He was worried that his family member who worked at University A might have access to the information. However, he grudgingly agreed after the researcher reassured him that his data would be kept confidential and could only be accessed by the research team and relevant regulatory authority.

Statement of the Issue

In this study—taking place in a multi-religious, Muslim-majority country—the religious, social, and legal implications of identifying a patient's MSM status requires careful consideration as a part of responsible conduct of research involving vulnerable populations (Barmania et al., 2016; Kanter et al., 2011). It is critical that the research team consider the potential consequences of a health-care provider performing informed consent in order to assure participants' privacy and averting unintentional fallout.

Outcomes

At the end of the study, as a token of appreciation, each participant was paid MYR50 for their time and effort, which they happily accepted. For claiming purpose, the university bursar required each participant to sign a receipt which included the project title, their name, and identification card number. The researcher, unsure if this was a breach of participant confidentiality, decided to comply with the university regulation that participants should provide this information.

Questions to Consider

1. Why are participants in this study considered vulnerable? Explain the ethical and legal decisions that revolve around conducting research in this population.
2. What are the practical implications of working with the HIV-positive and MSM populations in this study? How would this compare in your setting?
3. What could the research team have done to effectively recruit participants in this study?
4. In your setting, would there be other ways to provide a token of appreciation that would not involve participant identification?

Case Study Analysis

Review the Declaration of Helsinki (https://www.wma.net/policies-post/wma-declaration-of-helsinki-ethical-principles-for-medical-research-involving-human-subjects/) and the Belmont Report (https://www.hhs.gov/ohrp/regulations-and-policy/belmont-report/read-the-belmont-report/index.html). Which ethical principle(s) seems particularly relevant to you in analyzing case study #3? Write a brief "Statement of the Issue" paragraph that describes the ethical challenge(s) you identified.

Using your answers to the questions posed for case study #3 and awareness of ethical regulations and standards in your setting, write bullet points of the aspects you think should be included in the "Addressing Ethical Issues Involving Research Participants" paragraph of a case study analysis. If you are working in a team or classroom setting, compare and contrast your responses with those of your team or class.

Summary of Chapter

Men—particularly white men—have historically been the participants in clinical trials (Dresser, 1992; Konkel, 2015). This limited representation has meant that findings from some clinical trials may not accurately represent the sex- and gender-variation in pathophysiology of a particular disease (Regitz-Zagrosek & Seeland, 2012), or may not adequately address barriers to access or treatment experienced by patients who do not identify as men (van Diemen et al., 2021). However, men are also underrepresented in studies of certain diseases (Steinberg et al., 2021) and have a lower life expectancy than women (Lancet, 2019), highlighting that we still have work to do to improve men's health. Social stigma is identified in the case studies in this chapter as a contributor to factors that limit needed research on men's health, and the authors provide insight on ways to overcome these barriers.

The case study from the U.S. identified ways that public policy impacts what research can be done. Where potential harms may accrue to participants in terms of employment benefits or social circumstances, researchers may have to forgo research until social conditions are more favorable. The authors of the case analysis rightly point out that even when policy changes, there can still be social stigma. Thus, attention to protocol development that promotes participant safety is essential. They recommend attending to the concepts of autonomy (self-determination regarding study participation, and respect for potential participants' decisions) and confidentiality (security for enrolled participants' data) when developing study processes. In particular, they note that care is needed when combining clinical records with study data, due to the sensitivity of the information contained in clinical records. Relying on ethical principles can guide researchers as they develop these processes. Further, van Diemen et al. (2021) propose that stakeholders should be included in the development of study questions, protocols, and materials for recruitment and informed consent, as well as in identifying ways to reduce gender-specific barriers and burdens to study participation.

The case study from Malaysia reiterates these themes about how stigma can influence recruitment, enrollment, and ongoing study participation. In alignment with recommendations by van Diemen et al. (2021), the researchers relied on trusted clinical staff to recruit and enroll participants into their study. A second challenge arose when researchers wanted to provide a monetary token of appreciation for study participation but were worried that institutional requirements to collect personal information for payment might be of concern to participants whose sexual orientation was not yet public but could be inferred because they received money for a study of HIV in MSM. Possible solutions to this challenge are left to the reader to discuss, relying on the ethical principles outlined in the Declaration of Helsinki and the Belmont Report for formulation of ways a protocol could be revised to minimize privacy concerns.

The case study from India discussed attempts to include more geographically remote areas in research. The benefit to such an approach was inclusion of new sites, particularly those with higher burden of the disease being studied. However, laboratory analytics were required for eligibility assessment (HIV status), as well as monitoring the longitudinal health status of participants, and the laboratory quality in the geographically remote areas was of concern. Inaccurate laboratory data can jeopardize participant eligibility and health, as well as lead to erroneous findings that result from these data. The analysis of the case study offered numerous suggestions for ways to implement standardized protocols and quality assurance measures so that outlying laboratories could meet quality standards and contribute to the research. This attention to detail will allow laboratories in outlying areas to promote their quality and continue to meet standards for other clinical studies until such time as a national certification standard is available.

Overall, these case studies and analyses show how public perception and stigma, public policy, and research capacity are entwined. Where there is misalignment among these factors, researchers may have to either delay a research project in order to protect potential participants or develop and implement policies and procedures that will protect their privacy and confidentiality. In these case studies, the value of working with trusted partners in developing and conducting research was emphasized, particularly when the research topic is associated with stigma.

Chapter Exercises

1. For your research area, what political or social issues might impact participant recruitment and study implementation? Building from concepts of informed consent and protection of individual privacy and data security, describe safeguards you can build into your protocol to promote the success of your study.
2. VanDiemen et al. (2021) provide several suggestions for ways to improve inclusivity of participation in clinical trials. These include:

    - improving clinical trial design by addressing multiple factors that affect participation rates,*
    - increasing the diversity of the research team,
    - developing educational and recruitment materials that are appropriately sex- and gender-targeted [or other factors listed in the PROGRESS Plus* framework], and
    - improving access to clinical trial sites.

    [* One such framework is PROGRESS Plus, which recommends attending to these factors: **P**lace of residence, **R**ace, **O**ccupation, **G**ender, **R**eligion, **E**ducation, **S**ocioeconomic status, **S**ocial capital, and other context-specific factors (the "Plus").]

    For each of these bullet points, what steps can you take at your institution to achieve inclusivity regarding study participation?
3. Both *The Lancet* (2019) and *Scientific American* (2022) have identified reasons to improve inclusivity in study participation. In what ways are these ideas relevant to your research area? Consider how study findings may lead to practice recommendations and guidelines.

5 Language and Research Participation

The population eligible to participate in a given research study is usually determined by the focus of the research question and consists of people who have experienced and/or are at risk of experiencing either the outcome or exposure being studied. However, this requirement alone is not enough to ensure that populations for research are being selected ethically.

Researchers often find themselves surrounded by populations who do not speak the same languages as the research team. A part of Good Clinical Practice in the U.S., along with the ethics of research, is that a patient needs to be fully informed about the research in which they are being invited to participate. They need to be able to understand their rights to participate or decline and must not be coerced in any way. In order for this idea to be upheld, the research must be presented to participants in a language they understand. The U.S. has a federal guideline called the Common Rule that notes that information given to a potential participant about the research study should be provided in a language a participant understands (Resnik & Jones, 2006). Informed consent not only needs to be conducted in a language in which the participant is proficient, but any document discussing research should also be in that understood language.

International guidelines are less precise and may differ depending on the location. According to the Council for International Organizations of Medical Sciences (CIOMS, 2016), informed consent documents may not be required for participation; documentation of verbal consent done in the language a participant understands is enough for enrollment or declination. Canada has a similar policy to the U.S. Common Rule; however, it states that a language barrier should not limit an individual from participating in research and informed consent documents must be made available in other languages potential participants may need (Resnik & Jones, 2006).

Due to the extra steps and precautions needed to take place to include individuals who do not speak the main language of the research teams, i.e., having translators on hand, having documents available in different languages, some researchers might simply decide to exclude these individuals entirely. This exclusion poses several potential dilemmas to the integrity and generalizability of the research findings. It could potentially hinder the external validity of the results since they would only be

DOI: 10.4324/9781003033233-5

able to translate to those populations similar to the study participants. It also violates the directive in the Belmont Report to require fair procedures and outcomes for all participants (USDHHS, 2017). Rejecting and refusing participation in a study without any specific rationale is not considered ethical. This action adds to the issue of selection bias of participants and can skew results.

In addition to the written language for consent, researchers also need to take into account the literacy level of the populations they are trying to reach. Just because the consent or assent is written in a language that is understood by the participant does not mean they fully understanding what is written. Consent documents have language that is often complex and above the average participant reading level (Siegel, 2018). To help standardize the wording, institutional review boards (IRBs)/ research ethics boards (RECs) typically have templates researchers can use to guide them in writing research documents. However, in a study conducted in 2003, it was found that the standards given were still higher than the average reading level of the participants (Paasche-Orlow, Taylor, & Brancati, 2003). Literacy differs depending on age, even sex (Roser & Ortiz-Ospina, 2018). Thus, the language used, and time given when informing an individual about study processes must be adjusted accordingly. Time needed for understanding must also be given to the potential research participant in order to accurately determine if they understand what all is encompassed in the research, including possible compensation or rights outlined in the consent.

This chapter will provide case studies dealing with:

- language for youth with disabilities and their non-literate caregivers/family members;
- health literacy of human subjects when participating in a questionnaire about genomics; and
- clinical trials versus surgical intervention protocol descriptions when an injury occurs during a clinical trial.

Case Studies and Analyses

The case studies in this chapter are based in three countries: China, Morocco, and Switzerland. These case studies deal with how a caregiver does not understand the importance of an intervention and data collected about the impact of that intervention (Morocco), managing compensation for someone injured during a clinical trial and the language around insurance payment in a consent form (China), and the reticence of participants to answer a questionnaire when the vocabulary is too technical about genetics. Through the case studies, we come to understand (a) how important it is to start where the client (and participant) is in enrolling them into a clinical trial, (b) the need to be specific about compensation when a serious adverse event occurs, and (c) why translation into lay language can be crucial for obtaining information from participants in a social sciences study.

The case study analyses from three countries, the U.S., Germany, and India, highlight the regulatory differences that structure responses from cultural perspectives.

For the case analysis from the U.S., the U.S. Department of Human Services Common Rule is the regulatory basis for determining how to address consent form information and serious adverse events in clinical trials. In the U.S., IRBs expect study forms to be very specific in terms of how insurance payments are managed when an issue occurs in a clinical trial versus standard of care treatment in medical settings. In terms of child welfare, Germany clearly expects that the best interests of the child are managed from a systems perspective such that clinical trials are designed based on what is available in a country and how families are able to access resources. Finally, the case analysis from India underscores how technical language can ensure that data collection is not successful, especially in a region that has over 120 official languages with many not having the technical vocabulary for surveying people about genetics, etc.

Case Study #1 (China): Ethical Issues on Compensation for a Serious Adverse Event in a Clinical Trial (Adapted from Cong, 2018) by Yali Cong

Context

A 78-year-old woman came to hospital for knee replacement surgery in 2006. She was invited to join a global clinical trial, testing two drugs to prevent deep vein thrombosis (DVT): Xarelto® (rivaroxaban) or BAY 59-7939 (a factor Xa inhibitor). The knee replacement surgery was conducted on October 2006, and the patient experienced a serious adverse event (SAE) during the surgery. Symptoms included chest tightness, shortness of breath, palpitations, cough, sweating, a very weak pulse, blood pressure drop (to 60/40 mm Hg), and shock. The patient regained consciousness three hours after resuscitation. The hospital's principal investigator judged this complication to be a SAE and completed the SAE report accordingly.

Statement of the Issue

Although the SAE was related to a complication of surgery and not to the drug itself, it was an obvious direct consequence of participating in the trial. According to Chinese Good Clinical Practice, "the sponsor should provide insurance to those human subjects who participate in clinical trials, cover the cost of treatment and the corresponding economic compensation for the occurrence of the harm or death associated with the trial" (SFDA, 2003). A family member of the patient recalled that the informed consent form stated, "if a subject involved in this trial is injured during the study, the insurance company will pay correspondingly." Based on the consent form, and the principal investigator's explanation, it was clear that the trial sponsor had purchased global insurance for this clinical trial. Although the rescue cost was covered by the trial, the disagreement occurred when the patient and her family requested additional compensation from the hospital and the pharmaceutical company. After failure to agree on a compensation amount, the sponsor and the hospital were summoned to the Beijing Chaoyang Court by the plaintiff (the son of the injured patient) in 2008.

During court proceedings the insurance documentation was provided, showing that each participating individual's maximum insurance was approximately EUR 500,000 (Chao & Chu, 2009). As a final decision, the court required the company to pay EUR 50,000 for compensation. The plaintiff did not accept this amount and appealed to the Beijing Second Middle Level Court. This appeal was later rejected.

Outcomes

The lawsuit started in 2008 when the patient was 79 years old and ended in February 2013 when she was 85 years old. Her dreams of travel after the knee replacement surgery never came to fruition. Compensation of EUR 50,000 was paid directly by the sponsor, not by the insurance company. Thus, dealing with SAEs and compensation was dealt with internally, rather than through a formal procedure.

There are many implications from this case, including: the hospital Institutional Review Board (IRB) should have been aware of the review of insurance documents; the sponsor and hospital should have handled the payment between themselves and provided compensation to the patient without the need for the court's involvement; and the principal investigator should have understood the need to work with the IRB, the sponsor, and the hospital to manage this patient's compensation and care. One more point is the extreme vulnerability of an individual patient when they request assistance from an industry sponsor and/or a hospital. They are vulnerable relative to the cost of time, the cost to pay in advance for legal fees, the lack of support from hospitals, and even the physical and psychological damages suffered due to the long legal process.

Questions to Consider

1. How should sponsors participate in the process of informed consent for research studies and/or clinical trials they fund? What activities should they follow throughout the duration of the trial?
2. What verbiage should be included in the consent process to communicate with research participants so that responsibilities are clear in the event of a serious adverse event?

Case Study Analysis: A Perspective from the United States by Sadie Gabler

Context

This case study analysis reflects on the discussion of a clinical trial and serious adverse event (SAE) that occurred in China. There are a number of ways addressing the issues that arose from this case study would be handled differently in the U.S. context, such as: determining if the clinical trial experience can be linked to the SAE, how the clinical sponsor manages clinical care and associated costs that are not part of the trial, and how the informed consent form portrayed the clinical trial, including compensation for injuries sustained during participation. A detailed legal analysis of the contract language between the clinical sponsor and the investigative site may also be warranted.

Country Specific Regulations

This case study poses several ethical questions in addition to compensation and language for consequences associated with an SAE while participating in a clinical trial. In the U.S., commercial sponsors are not obligated to cover costs related to injuries that occur as part of research participation. The argument a researcher and sponsor from the U.S. would make would stem from whether the SAE was a direct result of participation in the clinical trial or inherent risk of the knee replacement surgery; in this case, both could be argued reasonably well.

As such, the risks associated with the standard of care treatments are not required to be included in the risk section of the research study informed consent unless there is overlap or an increase in likelihood or the severity of potential risk due to study participation. Therefore, the first potential issue that a researcher would need to address is what risks were listed in the informed consent form versus those that may not have been. Was the potential risk of a deep vein thrombosis (DVT) listed in the informed consent? If so, what was the likelihood of one happening in the target population? What were the known side effects of the two drugs being studied? Did the patient have a missed contraindication as part of the inclusion/exclusion criteria that would have excluded her from participating in the study? If so, then seeking additional compensation for harm would seem appropriate. Last, was this information delineated adequately in the informed consent, and was it written in lay terms to allow the average person to make an informed decision about whether participation was right for them? Research has documented that low health literacy levels impact individuals' understanding of and ability to make informed medical decisions in the U.S. and abroad, thereby impacting health outcomes. The National Library of Medicine's (NLM) definition of health literacy is: "The degree to which an individual has the capacity to obtain, process, and understand basic health information and services needed to make appropriate health decisions" (NLM, 2000). In the U.S., the average person reads and comprehends health information at about the fifth-grade level, with only 12 percent of the population being proficient in health literacy, as documented in the National Center for Education Statistics report in 2006 (NCES, 2006). Therefore, was the participant given the information in a language understandable to her as required by the Code of Federal Regulations, 21CFR 50.20, the regulation that governs the general requirements for informed consent (Common Rule, 2012)?

A researcher in the U.S. would be required to report this SAE to an Institutional Review Board (IRB) and the trial sponsor. The researcher would also be in contact with the study sponsor's medical director overseeing the trial to discuss the event and learn if other cases similar to this had been reported. Also, the local principal investigator for this trial would be responsible for determining the event's causality and whether, in their medical judgment, it was directly related to study participation. The fact that the statements (1) "the SAE was related to a complication of surgery and not to the drug itself" and (2) the complication was a "direct consequence of participating in the trial" is inconsistent from a Western perspective of direct consequence of participation. It would most likely be reported as a consequence of the standard of care treatment, knee replacement surgery, and not a direct consequence of taking part in the trial.

Unlike Chinese Good Clinical Practice, in the U.S., a participant's medical insurance would be responsible for covering the cost associated with their medical care and treatment from side effects of participation unless the individual's insurance provider clearly states in their policy that harm done as part of being involved in a research study is not a covered benefit. However, most insurance companies within the U.S. follow the Centers for Medicare and Medicaid Services (CMS) policy that states that they cover costs that arise from complications from participation in clinical research for reasonable and necessary care (CMS, 2007). Should the U.S. require sponsors to cover expenses associated with any harm caused by study participation? Although this may sound feasible, the costs associated with implementing such a requirement would directly impact the overall costs of health care and prescription medications, partly due to the potential for false claims, awards, and legal fees.

Addressing Ethical Issues Involving Research Participants

In this case, the participant and her family requested additional compensation for the harm caused by the SAE. The informed consent form stated that the sponsor had insurance to reimburse subjects should they become injured during participation. The question arises as to whether being in the study itself was a direct cause of harm to the participant or was the injury a result of possible underlying conditions and the surgery itself. This scenario begs the question, was any additional compensation justified in this case? In the U.S., consent forms routinely include subject injury language. This language commonly states that no additional compensation is available or set aside by any participating entities from which a participant can seek remuneration for time lost from work or for other damages as part of their participation. This type of language may dissuade participants from seeking compensation for injuries that may otherwise be questionable, unlike in the language provided in the Chinese case presented here.

In the U.S., clinical research subject injury language is a hotly debated topic at IRBs and during contract negotiations between research sites and sponsors. Many academic institutions are part of the federal or state government and may have immunity from lawsuits filed against them and their employees, including researchers. However, this immunity does not protect them from a case due to negligence on their behalf. Proving negligence on the part of the researcher, surgeon, or institution would be difficult. Therefore, the participant would be unlikely to receive any economic compensation for the harm caused by taking part in the research study. A case of this kind would be unlikely to make it to the judicial system in the U.S.

Case Study #2 (Morocco): To Treat or Not to Treat: A Case Study of Social Factors Impacting Clinical Decisions during a Pediatric Neuro-rehabilitation Program by Oumaima Zedgui and Mohammed Sbai

Context

Neurodevelopmental diseases are of growing prevalence being estimated at 420 cases per 100,000 inhabitants (Institut du Cerveau, n.d.). In countries where newborn delivery protocols are well followed and newborns are properly cared for, the prevalence is 1 to 3 children in 1,000 (Cans et al., 2008). In Morocco, 20 to 30 children out of 1,000 suffer from cerebral palsy (CP) (Personal communication with Professor Moulay Idriss Alaoui, 2020). CP is considered the leading cause of motor disability in Morocco. As in other middle-income countries, access to the Moroccan health-care system is still a challenge in some regions around the country due to different social, cultural, or/and economic difficulties. This issue is particularly the case for neurological rehabilitation. The Moulay Ali Institute for Rehabilitation (MAIR) is a not-for-profit outpatient clinic founded in 2015 to address this huge and tragic deficit. It is the first and only American-inspired neuro-rehabilitation facility in the entire country and serves the southern portion of Morocco from its clinic in Marrakesh.

MAIR treats patients from cities and rural areas, with the majority (60 percent) of patients being poor and unable to afford paying for services. Due to this, consent to treatments and research participation (i.e., clinical trials) needs to ensure that individuals do not feel coerced based on the compensation amount and afraid of research participation based on lack of clarity for the study portion of their interactions with the MAIR staff. MAIR therapists follow a U.S.-based proactive and comprehensive rehabilitation protocol. Their foremost goal is to trigger neuroplastic changes toward maximal recovery. Admission at MAIR starts by identifying neurological deficits and meeting with patients and their family members before starting the actual treatment. Therapy goals are set in coordination with patients and their family while carefully accessing their determination, motivation, and social support (Essery et al., 2017). Consequently, home therapy is a very important part of MAIR rehabilitation protocol and the MAIR clinic has been very interested in evaluating these practices to determine what makes an intervention successful for patients.

Statement of the Issue

We present here the case of a patient with severe CP caused by hypoxia during birth. He was admitted at MAIR clinic in March 2016 at age of eight. He came in on a wheelchair with crippling athetosis affecting all extremities. He also suffered from severe spasticity (1 on the Ashworth scale), had no head-held control, was unable to sit, crawl, roll over, or transfer. Coming from a poor family of six and divorced parents, the mother had no insurance and was the sole caregiver. MAIR offered a free comprehensive rehabilitation program that went on for three years. For all individuals who were children and/or who were unable to cognitively process treatment structures or answer evaluation questions, a legally authorized representative (LAR) was needed in order for them to participate in this program. The LAR agrees for the participant to partake in the program and clinical evaluation, agreeing to the procedures and compliance. The MAIR clinic staff provide verbal and written information about the treatment sessions and the need for the patient and LAR to answer questions evaluating their activities and satisfaction with the MAIR clinic's work.

The caregiver was committed to therapy sessions but never complied with home therapy. Nevertheless, the patient made significant progress and MAIR staff were very optimistic about his future. Unfortunately, the caregiver's behavior was not supportive as the patient's prognosis improved. She lacked candor and made deceptive statements regarding the home therapy and did not provide honest and accurate information when asked about the activities she was completing at home with her son. She also refused to participate in support group meetings. MAIR staff discovered that she was using the patient in a begging scheme on the streets and was using the MAIR therapy program to fraudulently obtain financial support from local NGOs and private donors. Therefore, the mother did not commit fully to the treatment options and did not provide accurate information for the evaluation component of this project.

Outcomes

Because of the overall attitude and behavior of the patient's caregiver throughout the rehabilitation program, progress was not optimal and MAIR staff feared the caregiver was determined to maintain/prolong all or some of the patient's disability status to stay in the program and receive the benefits of participation. MAIR clinic made the difficult decision to terminate therapy for the patient and their caregiver (the mother) and invest the limited clinic's resources in another patient who was fully committed to their recovery potential by adhering to participation practices and policies.

Poor socio-economic conditions combined with ignorance and illiteracy are a major hindrance to any progress in treatment and research (Essery et al., 2017). Communication is key to ensure the participant is aware of what ensures continued participation, as well as what procedures and compliance is needed. In this case, it prevented a promising human being from reaching his full potential in life. We believe this case is also a clear demonstration of the country's need for skilled social workers or individuals trained in the consent process to work with the MAIR clinic clientele to adequately explain the clinical trial and follow up at the clinic and at home for data collection. MAIR staff, who are not trained in research communications, tried their best in guiding the patient's caregiver, but they could not change her behavior and were victims of her repetitive deception. A skilled social worker would do a better assessment of patient's living conditions as well as family support at home. Better communication about program procedures and compliance requirements could have had different outcomes in this situation.

Questions to Consider

1. How can studies best balance research participation and patient best interests?
2. What language is needed during the consenting process about participation involving an LAR?
3. What balance is needed between patient welfare and patient participation? How should this be communicated to the participant and/or their LAR?

Case Study Analysis: A Perspective from Germany by Luisa Rusta

Context

This case describes a boy with CP, who was admitted to a clinical trial that included a form of monetary compensation. He was subsequently excluded from the trial because of his caregiver's non-compliance with study participation requirements. The researchers feared that the mother was determined to prolong her son's disease in order to stay in the trial for longer and, therefore, receive more compensation. The questions that come to mind about this case are:

- Was it ethical for the researchers to terminate the patient's participation in the study when he had shown that, with proper compliance, he could significantly benefit from it?
- Was the boy able to receive therapy outside of the research study?
- What was the reason for the mother's lack of compliance with study protocols?
- Furthermore, was it ethical to offer compensation dependent on the length of participation, keeping in mind that it could create a conflict between the patient's physical well-being interests and financial interests?

A good starting point is to look at the possible reasons for the caregiver's non-compliance with study requirements. The possibilities range from not having the proper resources for home therapy (or not even having a home) to illiteracy or poor health literacy to intentional negligence or abuse of the child. Because the mother's true intent is not entirely clear, the researcher's own judgment of the situation can be used as a basis for evaluating their decision.

According to the case study, the caregiver became less compliant as the patient improved, was repeatedly deceptive, refused to participate in support group meetings, and was found to be using the patient in a "begging scheme." It seems as though the researchers felt that the mother was intentionally preventing her son from improving too quickly. Whether she was concerned about being able to provide food, needed the money, or for other reasons is unknown. As a result, however, the data collected from the patient's participation was compromised and could not be used in the research. Therefore, as cited in the case, the patient was terminated from trial participation. If the researchers suspected ill-intent in the mother's actions, their decision to exclude the boy from their study without any further action was ethically questionable. First, it seems unfair for the boy to be punished for his mother's behavior. Second, abuse should be reported to proper authorities if such entities are available. Third, data collection from a patient should not be prioritized over the patient's well-being. On the one hand, it can be argued that because the patient was excluded from the study as soon as the data collected from his participation was compromised, he was being used as a means to an end, rather than an end of himself. On the other hand, the goal of the study is to collect data that can be used to improve the treatment for many other patients. Furthermore, I am unfamiliar with Moroccan laws and social services, such as child protective services, which the researchers could have relied on. It is possible they did not have the necessary infrastructure available to them in order to intervene effectively.

In the next section, I describe the steps researchers in Germany would have most likely taken in this situation and the resources that would have made these steps possible. I will also briefly describe how the other abovementioned reasons for poor trial compliance, such as illiteracy or general misunderstandings, would be handled in Germany.

Country-Specific Regulations

In the first part of this section, I will assume that the caregiver's action was coming from a place of negligence or abuse or was motivated by financial reasons. In the second part I will discuss the equally likely possibility that the researchers and mother did not find a common understanding about study participation requirements.

The German clinical research world today follows the core Kantian principle that humans are not to be treated as a means to an end but rather as an end in themselves. Based on Germany's history of human rights violations during the Third Reich and specifically the criminal conduct of many doctors under the Nazi regime, current German regulations for human involvement in clinical trials might be stricter than regulations in many other countries. Another aspect that sets Germany apart is the strong social support network that people living there can rely on.

In a situation where an underage patient initially benefited from a clinical trial, but their caregiver was found to be intentionally preventing their improvement, German researchers would make themselves punishable by law if they did not report the case to the *Jugendamt*, the equivalent of Child Protective Services in the U.S. The *Jugendamt* could then decide to put the child into medical foster care for the duration of their treatment or longer, if needed.

If we assume that in our case the continuous monetary compensation was contributing to the mother's actions, because of the family's poverty, then we can say that, overall, such a situation would be less likely in Germany. This situation is because of the many resources available to families, especially for *alleinerziehende Muetter*—single mothers. *Kindergeld*, for example, is money from the state each family receives for simply having a child, starting with 219 euros per child per month (Kindergeld, 2022). Additional support includes universal health care, rent assistance, and many other more specific programs.

If despite these resources, a family found themselves in poverty and was motivated by a study's monetary compensation, researchers could consult an ethics committee present in most German hospitals, consider changing the way participants are compensated, so as not to motivate them to prolong their disease, and rely on social workers to help monitor an underage participant's progress at home and help the family get the resources they need. In addition, to prevent unethical study designs, investigators must submit any consent forms, advertisements for the study, and monetary rewards to the independent ethics committee (the equivalent of an IRB in the U.S.) for review (ICHGCP, n.d.).

In the case where the underlying issue in the MAIR trial was a misunderstanding between the researchers and the caregiver, illiteracy or low health literacy could be two examples of possible communication barriers. Recent findings show that about two million people in Germany are fully illiterate and 7.5 million are functionally illiterate, meaning they are unable to easily read and understand more than one word at a time (Grotlüschen & Riekmann, 2011). Therefore, while the assumption of literacy is made in many legal interactions in Germany, there are a number of resources available to analphabetic people, including hotlines, social workers, and educational programs. The wide availability of social workers in hospitals can help prevent misunderstandings between clinicians and patients and their caregivers, as well as help with preparation for treatment, care during treatment, and rehabilitation afterwards (BKK-VBU-Meine Krankenkass, n.d.).

Addressing Ethical Issues Involving Research Participants

To review, some of the issues that could have played a role in this case study are how to communicate study participation expectations clearly to patients of all levels of education/health literacy, how to offer monetary compensation for study participation without creating a counterproductive incentive, and how to deal with potential cases of caregiver misconduct.

Ideally, study consent procedures should not be built on assumptions of health literacy or even literacy, because patients may not feel comfortable asking for assistance in understanding protocols and requirements, even if assistance is available on demand. Researchers should begin by establishing a potential participant's literacy and health literacy levels and based on that provide appropriately presented information about the study and its expectations.

A clear description of expectations and mention of federal/local law regarding a caregiver's responsibilities and the obligation of researchers to report any observed neglect/abuse could be helpful to include into a consent form when working with legally authorized representatives.

Should a study decide to use monetary compensation to help recruit participants, the compensation should be a set amount, be paid at the end of the study, and not be based on the health outcome of the patient or the duration of their treatment, so that patients have no incentive counteracting their interests of physical well-being.

Overall, this case study teaches us that clinical trial design is very much based on what resources are available to researchers in the particular country. Therefore, in addition to general legal and ethical standards, it is necessary to set country-specific expectations for the roles of researchers and participants during a clinical trial.

Case Study #3 (Switzerland): Lost in Translation: How Communication Affects Research Participation in a Multilingual Setting by Agata Ferretti and Caroline Brall

Context

Swiss researchers conducted a study about personalized health research and the opinions of the general public concerning the donation and use of their health data and genetic/biological samples for scientific research. The study consisted of a survey including 23 questions that was distributed to a stratified random sample of Swiss citizens, in both in online and paper-based formats. The participants varied in age, sex, gender, language spoken, education level, and socio-economic background. The survey touched upon a variety of topics using plain language but entailed medical-specific concepts. Initially developed in English, the questionnaire was translated by external translators into German, French, and Italian (the Swiss official national languages).

After conducting a pilot test, the researchers registered a low participation rate, and a high number of incomplete questionnaires. Some participants refrained from completing survey sections that they did not fully understand. Many participants were unfamiliar with the medical and statistical concepts used in the survey, and often misinterpreted the terminology used in the questions. For instance, many survey participants were perplexed when asked to complete sections about "personalized medicine," "likelihood to develop a disease," or whether they would prefer to have their health data stored in an "anonymized, coded or identifiable format." Furthermore, the questionnaire asked about participants' hypothetical willingness to donate their health data and genetic/biological samples. Despite mentioning the term "hypothetical," some participants were still concerned that they would thereby sign up to actually donate their health data and samples, and therefore refused to complete that section. Additionally, during the pilot test the researchers noticed that participants left some questions blank because the literal translation of the surveys from English largely diverged from what they originally intended to ask and occasionally did not deliver a meaningful message.

Statement of the Issue

In this context, the level of participants' medical literacy in combination with the language used to describe complex topics may have affected the participants' likeliness to fill in all survey questions. The technical, and potentially misleading, language used in the questionnaire might have affected study participation, introducing a selection bias which favors the most educated participants.

Moreover, the multilingual setting of Switzerland represented another challenge for the researchers: to conduct a survey study without accounting for the contextual linguistic peculiarities may have negatively influenced research participation. The researchers realized that to translate a survey in a variety of languages requires both high linguistic competence (to maintain linguistic equivalence across languages) and technical knowledge of the subject matter (to guarantee content equivalence across translations). Since not all languages have exactly the same expression to describe one concept, it is the researchers' duty to ensure that no biases are introduced in the translation.

Outcomes

After conducting the pilot test, the researchers—who were multilingual and at least two of them native in each national language—decided to translate the survey themselves anew. Translations were reviewed by two researchers to ensure that the content met their requirements for matching the intended meaning and understandability, thereby reducing the linguistic bias.

The researchers also refined the survey, taking into account the language and understandability issues. They clarified the questionnaire to clearly refer to hypothetical situations and simplify statistical concepts by rephrasing in plain language. To increase research participation, they also included an introductory section explaining that the survey's goal was not to assess participants' knowledge about the topics. Rather, the survey aimed at collecting participants' opinions on the matters.

Questions to Consider

1. In your context, would the language used affect research participation? If yes, how would you ensure that study participants meet the understandability threshold—both in terms of language proficiency and literacy of concepts?
2. Do you envision any concerns and risks of bias arising from the literacy disparity across study participants?
3. In the global research context, studies often involve multi-country partners, with a variety of languages/local dialects. How would you handle the translation or language adaptation to meet different local dialects and context in those research settings?

Case Study Analysis: A Perspective from India by Sarmishta Diraviam Kannan

Context

Conducting a study on the Indian subcontinent on personalized health research and the opinions of the public on the donation and use of their health data and genetic/biologic samples for scientific research would be a challenging feat. This secular country has 28 states, eight union territories (India Government, n.d.), and according to the 2011 Census 121 official languages and 270 mother tongues (India Census Bureau, 2011). In addition, the literacy rate (ability to read and write) of

India is 74 percent as per the 2011 Census report (India Census Bureau, 2011). Douglass et al. state:

> Unique to India, however, each of these regions is likely to have a different primary spoken language. Therefore, physicians in training in India are not only learning medicine, they are often learning the languages of their patients along the way. These factors result in a multilingual health care environment where ensuring language alignment presents a challenge.
>
> (2021)

It is evident that the number one hurdle in conducting a study on personalized health research in India is the language.

According to the 2011 Census report, there are six major religions practiced throughout India: Hinduism, Islam, Christianity, Sikhism, Buddhism, and Jainism (India Census Bureau, 2011). Religion influences people's thoughts, ideologies, and actions in their day-to-day life. Therefore, it is apparent that religion will also play a major role in the way the general public responds to a study on personalized medicine. It is important to factor the impact of religion while developing a study on this topic and is another challenge in conducting this type of study.

The wide range of socio-economic classes in India presents another challenge when conducting a study on personalized health. Social classification in India is determined by the per capita monthly income and is divided into six groups: upper high, high, upper middle, lower middle, poor and below poverty line. These divisions present an obvious disparity in access to education and health care, among other things, which would greatly affect a person's ability to understand and complete a survey about personalized health.

A study investigating the factors involved in reduced South Asian participation in clinical trials found that age, language, social class, culture, and religion are some of the top issues in research (Hussain-Gambles et al., 2004). In India specifically, the large number of different languages, lack of a national language, an average literacy rate, different religious affiliations, and lack of equal access to education, health care, and technology, are some of the major practical and ethical challenges faced by researchers trying to gather the Indian peoples' opinions on using their health data, genetic and biological samples for scientific research.

Country-Specific Regulations

Article 343 of the Constitution of India states that Hindi in the Devanagari script is the official language of the union; however, it is not widely accepted or acknowledged by the multilingual people of India, nor is it enforced by the Indian government (Indian Constitution., n.d.). An article in *The Times of India* newspaper from September 2021 reads that, even after 70 years, Indian politicians and the general public continue to fight over a national language. The article goes on to clarify that Hindi is not the national language of the country but only one of the many official languages (Choudhary & Sharma, 2021).

Thus, language stands as the primary barrier in conducting a survey on personalized medicine in India. Hindi is a language commonly spoken in the northern part of India in states like Uttar Pradesh or Bihar whereas Tamil is a language commonly spoken in the southern part of India in states like Tamil Nadu. According to Google Translate, the term "personalized health" transliterated in Tamil is "tanippatta arokiyam" whereas in Hindi it is "vyaktigat svaasthy." The same word sounds very different in these two languages which demonstrates the vast differences between languages across India. The same word in Kannada, the language spoken in Tamil Nadu's neighboring state of Karnataka is "vaiyaktikarisalagide arogya." Even languages from neighboring states in the same region of India are far from similar to each other. Adding to the language barrier are the multiple dialects of each language that are spoken in different areas within the same state. Not only do all the languages sound extremely different from each other, they each have their own unique script. This issue means that transliteration would not be useful, and the researchers would have to translate the survey into 121 different languages. That may seem simple with technologies like Google Translate; however, medical-specific terms may not be accurately captured by online translation services. A paper in *Nature* on the Indian initiatives to break science's language barrier states, "Few of India's languages have an up-to-date lexicon of scientific terms, and many researchers in the country have long become accustomed to thinking and writing about science in English" (Bharath, 2019). Therefore, even with the help of translation services and locals proficient in the language, it will be a daunting task to translate a survey on personalized medicine in 121 languages because of the lack of medical/science vocabulary in these languages.

Addressing Ethical Issues Involving Research Participants

Even if the survey was successfully translated, the study participants may not understand the concepts of personalized medicine or using one's health data or genetic/biological samples for medical research. This challenge is due to the lack of relevance of this issue in their communities and the level of education which can be attributed to their socio-economic status. People residing in rural areas may not have access to health care and therefore the idea of personalized medicine will be very new to them. In addition, the idea of using one's health data or biological samples for research will be a foreign concept and may be frightening to participants, which will prevent them from participating in such a study. As Chakrabarty et al. state in their perspectives of translational genomics paper:

> One of the major challenges in India is the lack of knowledge and general indifference in the majority of the population, including medical professionals, the public at large, and government bodies, to the potential benefits of application of genomics to the field of genetics for public health.
>
> (2016)

This challenge shows that there is a lack of understanding about personalized medicine even among the literate and medical community indicating that people able to read the survey translated into their language will not be able to understand these ideas to answer the questions posed. These challenges highlight that the study participants and results will not provide a good representation of the general Indian people's opinions on personalized health and using their health data for research purposes.

The case study provided also indicates that religion could not be distinguished from culture and cultural practices often caused people to shy away from participating in clinical trials. Another paper elaborates that religion can deter people from participating in trials where modesty was concerned as modesty is associated with religion for many South Asians (Hussain-Gambles et al., 2004). This idea can be extrapolated to participating in a survey about using biological and genetic sample for research. Some people may think that religion tells them their body is sacred and that using it for other purposes, like research, is taboo. These beliefs can influence their decision to participate in the survey and their responses to the questions.

Understanding the diversity in language, religion, and socio-economic status across India is vital to conducting a successful study on personalized medicine. Taking all these factors into consideration will help to provide results that are representative of the Indian population.

Summary of Chapter

This chapter highlights how impactful language use is and how illiteracy can have a detrimental effect on research in clinical trials and in social sciences studies. The issues considered in this chapter hinge on literacy but also on the language in which potential participants read and write. In countries where literacy rates are low in a spoken language (Morocco), participation and comprehension about what is expected and why is problematic. Along with this issue, countries where languages do not have the technical vocabulary (India) to convey information about a medical concept make it difficult for basic social science research to be completed—let alone more complex clinical trial information. A key issue in terms of language can also be about what is the motivation for participation—is it the compensation or the intervention/treatment services? Motivation will impact how participants and caregivers are able to provide adequate and accurate information about their activities in a study as well as how honest they are with the data they share. Language and literacy are separate concerns for any research study such that how information is provided can help or hinder data collection processes. Finally, consent forms need to be very clear about what is being offered to participants (China). For clinical trials, these forms should specify how injury will be managed in terms of care and cost reimbursements. On the face of it, all these issues would not seem to be connected; however, a participant's ability to read and/or understand what is being explained to them make these areas very much linked.

Chapter Exercises

1. Identify a basic survey or questionnaire. Determine if it has been translated into another language. Consider the questions asked. Ask a colleague who speaks another language how difficult would they be to translate into that language.
2. Interview a researcher in a biomedical field. How does literacy impact their research studies? If they conduct clinical trials, ask them how it impacts the data they collect for that trial.
3. Review your ethics board or IRB's website. What does it say about use of consent forms for clinical trials? Is there information about how to create a consent form for people with low literacy?

6 Participant Incentives

In many situations, it is necessary and appropriate to compensate research participants for their involvement in a study. Compensation comes in many forms, including food, gift cards, money, or free medical care, and is intended to thank participants for their time and effort. Research subjects spend their personal time participating in a study, and it is proper to provide compensation to limit the burden of the study activities on their personal lives. That said, the promise of compensation for a study should not influence individuals to participate if they would not have agreed to otherwise.

Overcompensating participants is known as "undue influence." As noted by one of the contributors to this chapter, "undue influence" and "coercion" are two different notions and should be recognized as such (Andorno, personal communication, 2022). The Belmont Report defines "coercion" as occurring "when an overt threat of harm is intentionally presented by one person to another in order to obtain compliance." By contrast, "undue influence" is defined as occurring "through an offer of an excessive, unwarranted, inappropriate, or improper reward or other overture to obtain compliance." The problem with undue influence is not that it involves coercion (the threat of harm), but that the lure of a profit distorts the freedom of consent. The Council for International Organizations of Medical Science (CIOMS) defines undue influence as "compensation [that is] ... so large as to induce potential participants to consent to participate in the research against their better judgment" (2016). Undue influence is considered unethical because it violates the concept of voluntary informed consent. The Nuremberg Code requires that participants should be able to "exercise free power of choice, without the intervention of any element of force, fraud, deceit, duress, over-reaching, or other ulterior form of constraint or coercion" when consenting to participate in research (Nuremberg Military Tribunals, 1949). If compensation for participation is too great, consent cannot be considered voluntary.

For low-resourced populations and settings, there is a particular risk for individuals consenting to participate in studies against their better judgment to collect the compensation. For instance, a study conducted in New York in 1951 rewarded homeless men with food, a place to sleep, and free medical care, in exchange for submitting to invasive and painful medical procedures (Aronowitz, 2014). The homeless population was chosen to participate in this study because no other men

DOI: 10.4324/9781003033233-6

would agree to such painful procedures unnecessarily. The amount and type of compensation imposed undue influence on the subjects in this study.

These concerns do not mean that research in low-resourced settings with underserved populations should be avoided altogether; it is crucial and ethical to include various populations in research, and they may be the most likely to need the benefits associated with research participation. The disadvantage some populations experience already places a disproportionate burden on them and some populations are less likely to have access to the knowledge and clinical benefits to which research is contributing. Underrepresentation in research prevents a population from enjoying the benefits of research, such as population-specific information about the treatment and diagnosis of disease. The CIOMS guidelines make a point to recognize that providing care to participants in low-resourced settings does not always constitute undue influence (2016). Protection of vulnerable populations from overrepresentation is achieved by requiring that the choice of research population must be "scientifically and ethically justified" and not based on convenience or "potentially discriminatory criteria" (CIOMS, 2016). The purpose of this guideline is twofold: to protect vulnerable populations from overrepresentation and/or underrepresentation in the research context.

Not least owing to the socio-economic and cultural variability in research populations, it is impossible to create regulations to set exact specifications for what sort of and how much compensation can be considered as unfairly influencing participants. Determining whether a study provides fair compensation is determined by individual research institutions. This issue poses a risk in low-resourced settings as the participant population in these areas is often vulnerable or in financial need, and institutions may not have the necessary regulatory infrastructure to prevent situations where there is risk for undue influence. Some of the issues the case studies in this chapter will explore are:

- what happens when community stakeholders who have been participating in a large research study request a renegotiation of incentives;
- research bias and justice issues in access to medications for participants recruited from private versus public sectors in a low-income setting;
- recruiting low-literate individuals and coercive actions; and
- research with former prisoner groups.

Case Studies and Analyses

The case studies in this chapter deal with research and providing incentives in four countries (Argentina, Ghana, South Africa, and the U.S.) are represented through these case studies. They deal with how literacy impacts people's ability to fully be informed about the parameters of a research or clinical study, how important it is for researchers to understand new needs of community stakeholders in developing and furthering a research project in the long-term, the need to distinguish between patients in a clinical setting when they cross over into becoming participants in clinical trial, and why it is crucial to have a clear understanding of policies for vulnerable populations or those populations susceptible to coercion.

For each of these case studies the concept of incentivizing participants or providing compensation for their time and knowledge is a key element of ethical importance.

The two case study analyses from Malawi and Nepal highlight the fact that there are no universal standards for determining how much compensation should be provided. For minimal risk studies, less compensation might be appropriate; however, for greater than minimal risk studies the incentives should not be so high that participants decide to take part in a study for the compensation alone. The need for considering participant stakeholders' voices in determining compensation is underscored in these case study analyses. In addition, the lack of standardized compensation packages is clearly described. It is worth noting that one of the case study analyses is specific about incentives, while the other case study analysis is very generic in this realm.

Case Study #1 (Ghana): Considerations in the Selection of Incentives for Community-based Participatory Research in Rural African Settings by Daniel Ansong and Scott Benson

Context

The University of Utah, Kwame Nkrumah University of Science and Technology, Komfo Anokye Teaching Hospital, traditional cultural community leaders, and locally elected government leadership in Ghana formed the Barekuma Community Collaborative Development Project (BCCDP) in 2004. This project enrolled approximately 20,000 community members across ten communities of diverse sizes into an ongoing community-based participatory research (CBPR) entity. Beginning in 2005, researchers, students and community members met on an annual basis to discuss scholarship activities that the two universities and hospital could conduct together with the community to teach students how to ethically conduct research with community partners. The community partners agreed to participate as long as they had a voice and would be informed of the findings of the research, as well as benefit directly from the research conducted. Ghana is considered a middle-income country in the context of health.

Over the course of 15-plus years, this ongoing project has produced multiple advances for the health of the communities including better understanding of malaria in children under five years of age, a significant reduction in the burden of schistosomiasis in children, the identification of a silent epidemic of hypertension in adults (i.e., with >40 percent of adults over 18 years of age having undiagnosed or untreated hypertension), as well as multiple water, sanitation, and hygiene (WASH) projects.

Statement of the Issue

The BCCDP is situated in a peri-urban region of the Asante region of Ghana, just outside of Kumasi, and is considered primarily a rural setting. These communities face challenges with access to clean water and sanitation. Study participation was frequently incentivized by the provision of a bar a soap in exchange for community members' participation. Soap was a commodity selected for the incentive as it was not universally available in the communities. In addition, its use improves health, it does not spoil or expire, it could be easily transported and exchanged for participation, and, hopefully, its use induced good health behavior, regardless of the focus of the study.

Typically, the soap was well received and accepted as a reasonable incentive. However, starting in 2012, commercial trucking increased the amount of product availability in some of the larger communities, specifically those geographically closer to the city of Kumasi. In these communities, soap was now readily available and community members wanted to discuss alternative incentives. The research team met with local academics, tribal and governmental leadership, and community members to discuss this concern. As part of that discussion, it was felt reasonable to offer different incentives within different communities depending on local needs. However, the equivalency of cost became a sticking point. The discussion focused on the difference between cost and value. Some team members insisted that each individual who participated should receive an incentive of equal cost, though the relative value or benefit to the individual may be different. Other team members were more concerned that incentives advance locally specific health outcomes of perceived equivalence, regardless of cost differences. As an example, a larger community, where some community members expressed their dissatisfaction with bars of soap, stated they wanted the construction of a malaria research center for the community.

Outcomes

In high-income countries, the debate regarding provision of incentives has become simplified with gift cards of common-equivalent value to stores, such as Amazon, where there is a wide variety of products available for purchase. While these options will eventually reach rural low- and middle-income countries (LMICs), such options are less available at present and the conundrum of appropriate incentives for participation in minimal risk research studies continues to be a challenge for researchers, communities, and participants.

Questions to Consider

1. For equivalent participation in research, should incentives be tied to local economic value which might vary across the geographic region of a study?
2. Should researchers be selecting incentives based on a perceived beneficial use that is unrelated to the study at hand or any expertise of the research (i.e., soap for participation in diabetes surveillance)?
3. Who decides the appropriate balance of cost equivalence for equal participation versus value/benefit equivalence for equal participation?

Case Study Analysis: A Perspective from Malawi by Yohannie Mlombe

Context

At the current time, there are no specific guidelines for the payment of research participants in Malawi. Investigators usually propose their own amounts with different research institutions proposing different fixed rates for what was called "transport reimbursement" for participants. These amounts range from the equivalent of $1US per day (about MWK 1,000) to $20US per day. The most popular amounts are between $1US per day and $5US per day. Ethics committees in Malawi tend to ask investigators to revise down any participant compensation amounts that were more than $5US per day on the basis that such amounts would unduly influence study participants. The prevailing thinking is that because the minimum wage in Malawi has remained around $1US per day, any participant payment above this is too much or could be undue influence.

Country-Specific Regulations

In Malawi, it is not uncommon for investigators to argue that their research involves concurrent review by external/international ethics committee/IRB or is funded by external/international institutions that do not allow participant compensation or would not approve amounts recommended by local Malawian IRBs. When specific ethics committees/IRBs/institutions have been provided and one bothers to check their guidelines, almost always one finds no specific written guidelines from those ethics committees/IRBs/institutions stating what is prohibited. In fact, one finds guidelines that recommend higher participant compensation amounts or one finds publications that were approved by the named entities with higher participant compensation amounts than those being recommended by local Malawian IRBs.

Payments to study participants should consist of distinct components that include reimbursement for expenses incurred by the study participants, compensation for time and burden, and incentives. A meeting held in Malawi in 2018 sought to balance these components and make appropriate recommendations. Activities such as transport rates, time, and burden categories were used to calculate a strategy for minimum remuneration for all participants. For transport rate, it was recommended that researchers use the same amount for all participants in the study and base it on furthest travel distance, i.e., 5km = 300MK, 10km = 600MK, 15km = 900MK, etc. For time, research should use the total time including travel, waiting, consultation, tests, waiting for results, and treatment, etc. In terms of burden categories for medical studies, compensation should be considered by discomfort levels. Finally, it was determined that all researchers should understand there would be a minimum remuneration such that compensation should not be less than 7,000MK for certain types of greater than minimal risk studies.

Adopting the above concepts in Malawi is helpful, but it has its own issues. The obvious criticism is that study participants were not involved in the formulation of this guideline. Ethical guidelines should also be clear enough so that researchers can budget accurately ahead of their protocol submission to an ethics committee, i.e., when they are submitting grant proposals for funding. A complex approach raises the likelihood that participant payments may vary widely and be unpredictable when a study protocol is submitted to an ethics committee since investigators and ethics committee/IRB members may weigh participant time, costs, and burden differently. As such there is merit in fixing a recommended national participant payment amount in Malawi and vary that amount based on specifics of the study using the guideline, which is available at https://wellcomeopenresearch.org/articles/3-141/v2?src=rss

Addressing Ethical Issues Involving Research Participants

Study participant payments are a controversial subject in many countries because there may be a "hidden" conflict of interest. In research protocols, investigators and sponsors of research may not routinely state their own vested interests in the research that they are doing. Research protocols highlight benefits to society, policy development, patients, etc. While research benefits society, policy development, and patients, research also substantially provides financial and academic gains to sponsors and investigators. It is fair to say that without some type of personal or financial benefit, very few investigators would conduct research. In fact, it can be argued that the result of most applied research is the discovery of patentable products and subsequent commercialization. Therefore, to expect research participants to bear the burden for the advancement of society for nothing while investigators and sponsors reap the benefits of research is unethical. There is a power differential between investigators and research participants, which is usually most pronounced in low-income countries, like Malawi. In low-income settings, career and/or academic researchers can be among the richest individuals. Ethics should mainly be about balancing power differentials—as noted by Mikesic, "I'm trying to imagine ethics without an awareness of power. That would be like trying not to step on anyone's toes, without an awareness of one's feet" (Barstow, 2020). As an illustration of this power differential, an investigator may argue that a research participant who has reported to a hospital for routine care should not be reimbursed transport because they were already coming to the hospital anyway. But the same investigator may be receiving extra salary for providing standard of care in both the medical and research umbrellas—thus, they may be receiving double payment for everyday work and research activities.

For investigators, paying study participants may mean a reduced budget for their own payments from the research, representing a clear conflict of interest. Research ethics committees require a strong ethical stance to act for research participants beyond the *primum non nocere* ("do no harm") concept. Based on work by Mduluza et al., "Some of the pertinent questions by the research participants that go unanswered include the following: i). What is my immediate benefit? ii). Who will benefit from this work being conducted? iii). Why are these people (researchers) using all the resources in my community and yet we have so many other problems? iv). Why my community and not that other community? v). The researchers are here for only 2 years, then what next? vi). These people are well off, better than anyone in our community, so they must be benefiting from the activities?" (2013). If these types of questions continue to go unanswered, it may eventually become difficult to recruit research participants in resource limited countries like Malawi. As Cheff noted,

> "If you start with the assumption that their time is as valuable as yours then it changes how you think of compensation . . . I think of honorariums as compensation for people's time. If I am getting paid, they probably should be too. And I prefer to provide cash so that participants can decide how they want to use the money . . . I think it is morally wrong that we give participants grocery vouchers instead of money. We want to dictate to them what they should spend the money on when it is their choice how they want to spend it"
>
> (2018)

Additional considerations for compensation for participants are items such as what is the minimum wage and what type of research is being conducted (e.g., social science research, observational research, clinical research). Participant compensation for their time has often been linked to minimum wage; however, a more ethical consideration is the concept of a "living wage." Sub-Saharan Africa and South Asia are now home to over 80 percent of the world's poorest as classified by the World Bank, i.e., individuals living on under $1.90US per day (ShareAction, 2021). In Malawi as of February 2019, the minimum wage was below $1.90US per day.

In conclusion, incentivization would imply payments that motivate individuals to participate in a study. The general argument by investigators is that in a low-income country, like Malawi, any payment to study participants, even a genuine compensation for their time, would likely be an incentive that would be undue influence. This argument seems unethical in that it appears to take advantage of low research population's situations to deny them what they should rightfully and fairly earn. An incentive should ideally be payment to the study participant that is above the total of reimbursement, time compensation, and burden compensation.

Case Study #2 (South Africa): Low-Income Participation: Vulnerable and Ethical Considerations by Hannah Simonds

Context

Many international multi-site studies include low-income countries as participating sites. In particular, oncology clinical trials using newer technologies with immunotherapy are open to recruitment in many trial centers including patients from the public or state-funded sectors, and those who have private medical insurer funding.

For example, a study of immunotherapy in metastatic lung cancer is commenced. The study allows for limited post-trial access if the outcomes show benefit and allows continuation of the study drug until it becomes commercially available in the U.S. Participants, who were patients, were recruited from three public sites and three private practice sites in a low-income country. The patients in the private sites have medical insurer funding and could receive second-line therapy with additional targeted agents, upon disease progression. The public sector patients cannot receive second-line therapy and will receive supportive care only.

Statement of the Issue

Clinical trials recruiting vulnerable patients from low-income settings provide expensive oncological therapies only within the study time period and for a short period of post-trial access. For low-income participants, these agents will not be available once registered for commercial use, nor will they be available for the broader public patient group in future. The private sector patients, who can most likely pay out-of-pocket for their health care, will have access to this drug, and others, if needed. The clinical research team had decided to use the drug and standard of care for the incentive; however, this decision set up an unequal compensation model for the private versus public sector participants.

Outcomes

The ethics committee asked that the sponsor provide assurance that the drug would be available to study participants in the public sector even if it becomes commercially available to private sector patients. This request did not address the issue of second-line treatment or access for public sector patients whether they participated in the research or not. In addition, the differential compensation model was never discussed.

Questions to Consider

1. What are the ethical obligations of pharmaceutical companies to the patient populations that are recruited into their studies when these populations come from vulnerable, low-income sectors of society in the low-income countries?
2. Should this obligation include ongoing access to study drug post-marketing?
3. Should this obligation include compassionate access for other individuals not recruited to these studies?
4. What incentives should have been offered? Why those incentives?
5. What are the issues for the Belmont Report's principle of justice (equal distribution of benefits and harms) that the ethics review board might address in this example?

Case Study Analysis: A Perspective from Nepal by Gehanath Baral

Context

Biomedical research studies are expensive and require strict regulation to protect the rights of research participants at a study site. Research regulations may not be well developed or are in an initial phase in low-income countries. Research culture as well as the expertise in some of these countries may be underdeveloped due to lack of education on bioethics and research topics. Persuading authorities and stakeholders about the value of research would be the first barrier to overcome. The second barrier would be to help people understand the importance of research especially if there is no direct advantage to them when they may have some degree of basic health need. If the research outcome is not reasonably beneficial to the population at the study site in the future, a study could be mere exploitation of the unprotected people. For example, if an expensive anticancer drug is on trial and that drug would not be available at an affordable cost to that population even if the outcome were good, that type of research is not justifiable. In addition, if participants are not offered equal compensation for the same research activities as participants in other countries, a serious issue of ethical justice arises.

The strength of ethical committees and technical reviewers on these committees may be at a beginner level, which may adversely impact participants' rights as well as the quality of research. There may be risk of bias in the selection of participants and risk of inconclusive outcome despite reasonable efforts to avoid these biases; thence a participant may take risks unknowingly. The lack of clear understanding of ethical principles and their inappropriate application during ethical review process could be detrimental. The hierarchical relation of research staff to the research participants, lack of controlling for confounders, incoherent research design and objectives, ill-defined sampling frame and conceptual framework, and use of non-validated research tools are areas that should be addressed by the global research community.

Country-Specific Regulations

Many national regulatory authorities develop their research guidelines based on the Helsinki Declaration, Good Clinical Practice guidelines, and other international documents; however, there may be some challenges to fitting this information into geographical and cultural context. In addition, even when these guidelines are in place, particular authority if politically appointed in a regulatory post may disregard the expertise of an ethics review committee that is supposed to be in a scientific decision-making position. Let's examine two scenarios, for instance. In the first scenario, there was pressure and administrative power exercised on an ethical board for a clinical trial that was not well versed on beneficence and justice to the participant and the society. An educated and experienced ethical board could sustain tremendous pressure for a long period of time from the administrative power. In scenario two, statute and regulations were deviated by bypassing the scientific authority for the sole benefit of commercial stakeholders. For the first case scenario, the research proposal appears to have been reviewed by subject experts and appropriate ethical bodies, but the principles of beneficence and justice are potentially compromised if not handled independently. For the second scenario there is a loophole created by a corrupt factor that tends to inappropriately legalize the condition.

Addressing Ethical Issues Involving Research Participants

Awareness training, lobbying and scientific capacity building could be a solution. There should be a mechanism to keep the ethics committee independent with full scientific authority. There should be a structured program for training technical reviewers in the review process, the researchers about ethical principles and how to apply them in appropriate research design, good clinical/manufacturing/laboratory practices, and relation of research to people and society. Awareness and orientation programs will be necessary for community leaders and authorities. Enabling researchers, sponsors, and regulatory authorities to be empowered in the areas of beneficence and justice would be important to manage participants' rights and societal need. In addition, incorporating research ethics and moral values in school/college curricula is crucial.

Case Study #3 (United States): Compensation for Vulnerable Populations: Navigating Participant Access, Rights, and Privacy by Jordan Johnson, Ida Tovar, and Katelyn O'Farrell

Context

In the U.S., drug courts are a therapeutic jurisprudence program designed to recognize and rehabilitate persons with substance use disorders to break cycles of criminal behavior, offer problem-solving support, and eliminate drug dependence (Logan & Link, 2019). Stigma against individuals with a known substance use disorder and/or criminal history has the potential to increase maladaptive behaviors (i.e., secrecy and isolation) and negatively impact quality of care (McCann & Lubman, 2018).

At a university hospital in the U.S., researchers were examining the stigma drug court participants faced when receiving services or medical care. The study was conducted in partnership with the local drug court program and its clients in pursuance of detailed experiences. Adult drug court participants were recruited through flyers dispersed by drug court administrators and on Facebook, inviting them to share their experiences. Participants completed a one-time, semi-structured interview with members of a trained, qualitative research team. Participants were asked questions about drug court, health care, housing, employment, stigma and perceptions of addiction, and the COVID-19 pandemic. Following the interview, participants were eligible for reasonable compensation via gift card at a local shop.

Under the U.S. Office of Human Research Protections, 45 CFR 46.303(c) (OHRP, n.d.), prisoners, including those detained in treatment facilities, are protected against unethical or harmful research protocols. However, it was deemed that the drug court participants in the present study were distinct from this population because they "voluntarily" enrolled in the drug court program and were living in their own communities, rather than being detained. These individuals were recruited to the study via flyers and Facebook advertisement, with compensation promised for participation. These participants were, however, under contract with the drug court to comply with program policies and procedures (Overson Law, PLLC, n.d.). As a "non-prisoner"—yet—susceptible population, individuals who agreed to participate in the research study had fewer specific protocols to safeguard their enrollment. In addition, the impact of COVID-19 on conducting in-person interviews played a role in confusing how incentives could be provided. Given this unique exposure, drug court administrators became more conscientious and protective of drug court participants.

Statement of the Issue

In this research context, providing compensation to drug court participants could be seen as a form of coercion due to the link to recruitment and incentive provisioning. Given the nature of their engagement in the drug court, the target population is susceptible and at an increased risk for negative impacts associated with being identified in a research study. The COVID-19 pandemic presented an additional layer of complications. COVID-related research precautions required remote research participation, including compensation methods. These issues presented a challenge for incentivization, given a policy from the research institution's accounting office requiring a physical signature from each participant confirming receipt of compensation.

Outcomes

Innovative and strategic adaptations were needed to address the issues mentioned above. To reduce the risk of coercion, participants were told their participation in the research was voluntary and would not affect their standing in the drug court in any way. The compensation amount was also set at an amount deemed reasonable for the time and effort put forth by participants. To protect the identity of the participants, document compensation distribution, and practice safe COVID-19 protocols, gift cards were sent electronically and the email correspondence between the participants and the research staff were saved. The emails were kept separate from participant study IDs and were only used for financial auditing purposes, ensuring that in data analysis and paper preparation participant identities remained anonymous. Participants were also given the choice to opt out of compensation, but still participate in the study, if they did not want to have any of their identifying information stored.

Questions to Consider

1. What are the susceptible/vulnerable population rules that should be considered in your research setting for prisoners, people on probation, or mandated to a court program?
2. How do identifying factors (i.e., drug court participation) impact issues around de-identification, consent, incentives, etc.?

Case Study #4 (Argentina): The COMPAS Case: Exploitation, Informed Consent, Trust and Illiteracy in Argentina by Felicitas Holzer, Alahí Bianchini, and Roberto Andorno

Context

In Argentina, in August 2005, a large pharmaceutical company based in the U.S. was conducting a pneumococcal vaccine study on nearly 14,000 children in the provinces of Córdoba, Mendoza, San Juan, and Santiago del Estero (Homedes & Ugalde, 2014; Tealdi, 2015). As the biomedical ethicists, Homedes and Ugalde (2014) write, "The Governor declared the vaccination with the experimental pneumococcal vaccine a Policy of the State. For this reason, physicians in the public facilities would recruit patients" (Calvo, 2009).

According to the authors, "many parents of the children participating in the trial did not understand the scientific terms contained in the 13-page informed consent form" (Homedes & Ugalde, 2014). Furthermore, Tealdi (2015) reports that ethical problems not only concerned the informed consent process of trial subjects but also the non-compliance with the inclusion criteria of participants, concerning their age.

Statement of the Issue

It was generally assumed that people using the public health facilities in Santiago del Estero had confidence in the physicians and the care facilities. Thus, recruitment was fairly easy and only 14 parents who were invited to participate with their children declined (Homedes & Ugalde, 2014). However, due to high illiteracy rates, parents did not appropriately understand the scientific terms contained in the lengthy, high-language-level, technical informed consent form. For the parents, their incentive to participate was to access care, although, in many cases they did not understand that they were undergoing a clinical trial, so we could say that they were not adequately informed and, therefore, were not able to consent and were not provided free choice. According to the Argentine newspaper *Página*12 (2007), the President of the Argentine Federation of Health Professionals (Freposa), Jorge Yabkowski, had stated that researchers would "take advantage of poor mothers who are not told that their children are going to be subjected to a protocol, make them sign without reading and even threaten them if they want to abandon the study."

Following Ugalde and Homedes (2013), the sponsors of the COMPAS trial received an administrative fine from the Argentine National Agency of Drugs, Foods and Medical Devices (Administración Nacional de Medicamentos, Alimentos y Tecnología Médica, ANMAT) for serious violations not only in relation to informed consent but also in relation to the inclusion/exclusion criteria required by the trial protocol and, in the authors' words, the double vulnerability of participants.

Outcome

In the overall study in Argentina, Panama, and Colombia, 26 children enrolled in the COMPAS trial died. Homedes and Ugalde (2014) report what happened as a result of the case. In the authors' words, "seven of the Argentine children had been recruited in Santiago del Estero, while the other deaths occurred in the provinces of San Juan and Mendoza" (Homedes & Ugalde, 2014). A relationship between the deaths and the vaccine was denied, as the children enrolled in the trial had overall lower mortality rates than the average infant mortality rate in the respective countries and provinces. Nevertheless, the sponsor paid compensation to the families who lost their children without accepting responsibility for the deaths. A subsequent inspection by the Argentine regulatory agency (ANMAT) found that children did not receive the examinations specified in the protocol and that some did not even meet the selection criteria for study participation. Furthermore, the sponsor provided assurances that 98 percent of the parents or custodians received education, among which 70 percent showed a higher education level, suggesting that parents were able to give informed consent. However, the ANMAT inspectors stated that many parents and custodians of the research subjects signed the informed consent form without understanding the objectives and the conditions of the study. In the case of illiterate patients, informed consent was not obtained in the presence of two witnesses. The government put forward a formal complaint after the ANMAT inspection incurring a fine of one million pesos (around $320,000US in 2009), a rather symbolic amount of money, to the responsible investigator and coordinator in Argentina, to the principal investigator in Santiago del Estero, and to the pharmaceutical company. Among the reasons stated, the responsible actors were accused of failing to implement the necessary actions to comply with good clinical practices in the development of the clinical trial, and for having poorly carried out the process of informed consent going along with the poor quality of information given to parents about the scope and risks of the study. Incentives to participate in the study were set on the basis of misinformation and threats without any true compensation for time and effort on the study.

Questions to Consider

1. What requirements need to be met before conducting a study to assure that (a) the study is appropriately conducted following a protocol, (b) information about the study is appropriately disclosed to participants, and (c) incentives to participate in the study do not amount to an "undue inducement"?
2. What ethical review and institutional oversight mechanisms are available in your research context to detect ethical problems in the informed consent process of a large-scale clinical trial? Once these morally problematic aspects are detected, what bodies are assigned to solve them?
3. What institutional oversight mechanisms exist to ensure appropriate informed consent procedures when working with vulnerable populations?
4. What inclusion and exclusion criteria should exist for a clinical trial to be considered equitable?

Summary of Chapter

Determining when or to what extent compensation for participation in a research study or clinical trial is ethically acceptable can be quite complex. At the current time, there are no universal guidelines for how to manage this issue and be equitable in providing incentives to participants or ensuring that the time they spend on study activities is recognized. In addition, there are questions about how much to provide for participants and what are the parameters for doing so. Should compensation for participation in minimal risk studies be less than for greater than minimal risk or clinical studies when the time spent participating is the same, but the activities are different? What should participants be compensated for, i.e., travel logistics, time, childcare, etc.? How should the compensation be provided, e.g., gifts cards, online payments, food, soap, etc.?

An interesting consideration is that the word "incentive" means to "motivate or encourage one to do something"; while the word "compensate" means to "give something, typically money, in recognition of loss, suffering, or injury incurred" (*Oxford English Dictionary*). These definitions set up a quite different understanding about interactions with participants and how to ensure that their involvement in the study is acceptable. That said, these case studies raise questions about how to include participant stakeholders' voices in determining what type of and how much compensation they would find appropriate for their time in a research study. The issues surrounding compensation and incentives need considerably more research so that a more universal standard could be applied both to international research and to country-specific research.

Chapter Exercises

1. Review the case studies described in this chapter, with a particular focus on the compensation aspect. Develop your own standard guidelines about what compensation or incentive should be provided and why. Create a comparison table of the case studies that would highlight the elements important when considering compensation or incentives in those settings.
2. For one of the case studies without an analysis, draft your own analysis. Outline what would be the major considerations for providing incentives to participants. Why?
3. In some studies, especially clinical trials, a principal investigator and members of the study team may be provided an additional monetary incentive to conduct the study. What are the ethical issues when this information is not disclosed to participants? What information should participants know about this type of compensation?
4. Consider how you might include community stakeholder voices if you are developing compensation or incentive guidelines for a research study or clinical trial. Which voices should be included and why?

7 Mental Health

People with mental illness and/or cognitive disabilities may be participants in research studies. According to the Nuremberg Code, in order for voluntary informed consent to be valid, "the person involved must have the legal capacity to give consent" (Nuremberg Code, n.d.). Depending on the nature of the illness, some individuals with mental health issues may not have the ability necessary to legally consent for themselves. There is no global consensus on the best method of determining whether a person is capable of consenting for themselves, as the concept of self-governance can be situational (CIOMS, 2016). Some people may be perfectly capable in the morning and become incapable by the afternoon, while others may be capable of making certain decisions autonomously (CIOMS, 2016). Determining the parameters for participation for these individuals is important so that vulnerable populations are included in research and they are adequately informed about what participation means to them.

CIOMS Guideline #16 addresses how to ethically obtain consent from adults who are deemed incapable of consenting for themselves, i.e., the potential participants (2016). In cases like these, participation follows a two-step process. First, the responsibility for consent falls to a legally authorized representative (LAR), a competent adult who has legal authority to care for the potential participant. Second, the potential participant must still provide assent to participate, defined as more than "just the absence of dissent" (CIOMS, 2016). The process of assent requires that the researcher engage the potential participant, at an appropriate cognitive level, in discussion about the research purpose and expectations, allowing opportunity for the potential participant to assent or dissent to the research. The dissent should only be overruled in circumstances where the treatment offered is only available in the research context and both the legally authorized representative and treating physician agree that the treatment is in the best interest of the potential participant. Care must be taken during the assent process to ensure that the circumstances surrounding sharing information and answering questions are not coercive.

Research involving participants with mental illness must navigate the fine line between protecting a vulnerable population and excluding that population from important research. Unnecessary exclusion that results from researchers overprotecting a group with a particular condition (such as mental illness) is a form

DOI: 10.4324/9781003033233-7

of discrimination, which can increase stigmatization and restrict advocacy for the overprotected population; further, exclusion could result in a research gap and deficiencies in evidence for best practices for the care of people with the defined condition (APA, 2006). This selective exclusion is especially detrimental for those with mental illness because this condition tends to occur with many comorbidities, complicating care and emphasizing the need for thorough research that leads to a better understanding of research interventions on both mental and overall health (APA, 2006).

The historical unethical overutilization of certain groups, including those with mental illness or who are unable to consent for themselves, has resulted in the development of additional protections for these groups, as found in the U.S. Department of Health and Human Services regulations for human research subjects with vulnerable populations (OHRP, n.d.). However, the CIOMS guidelines aim to prevent populations from being "categorically excluded" from research, noting that their inclusion may require additional effort in the consent and approval processes, as well as during conduct of a study (CIOMS, 2016).

The case studies in this chapter consider the following topic areas:

- balancing issues of convenience, benefits, and risks when presenting research to a potential participant who may experience significant challenges to participation;
- protecting participants' privacy (to avoid negative employment consequences when research is conducted in the workplace) while addressing concerning findings; and
- managing research activities when participants are potentially stigmatized by additional factors (in this case, sexual orientation or gender identity), i.e., intersectionality (the ways in which different identities—such as race, class, and gender—intersect to produce individualized experiences of discrimination) (Coaston, 2019).

Case Studies and Analyses

The case studies in this chapter are based in three countries: India, Turkey, and the U.S. They deal with the need to emphasize both benefits and risks, and identify ways to facilitate involvement, when presenting research opportunities to participants at risk for mental distress related to the combined burdens of participation and ongoing physical health challenges (India); developing mechanisms to protect research participants' privacy while still addressing concerning findings among workers who would benefit from referral to mental health services (Turkey); and finding creative ways to meet participants' mental health-care needs who have intersectional identities of poverty, rurality, and LGBT status (U.S.). Through these case studies, we see how (1) participants with multiple comorbidities can experience burden to participation that are more readily overcome when the benefits of participation are clearly articulated, (b) research in occupational settings raises additional concerns for personal privacy and data confidentiality when research findings may impact employability, and (c) stigma associated with intersectional identities may create access barriers for people seeking treatment for mental health issues.

The case study analyses are provided by researchers in three countries: U.K., India, and Hungary. The case study analysis from the U.K. points out the importance of building in additional protections for participants who will likely experience distress during research through the design of recruitment and informed consent processes, and the research protocol. The authors recommend the use of patient and public involvement (Cochrane Training, 2019) as a method for identifying and addressing participants' potential concerns when designing the study protocol and preparing for ethics review. Similarly, the analysis from India highlights the need for exploring potential outcomes of participation, setting expectations for likely scenarios that relate to outcomes and risks of participation, as well as benefits, so potential participants can make an informed choice about enrolling in the study. The analysis from Hungary places intersectional vulnerabilities within the broader sociopolitical context, demonstrating that approaches to providing mental health care and to promoting study involvement can be highly context specific.

Case Study #1 (India): Clinical Social Work Intervention Research for Persons with End Stage Renal Disease in India by Anish K.R. and Ginu George Powathil

Context

A study among persons with end-stage renal disease was planned to reduce the psychosocial impact of long-term illness. Globally, the number of individuals with chronic kidney disease (CKD) at any stage reached almost 700 million in 2017, which is more people than those with diabetes, osteoarthritis, chronic obstructive pulmonary disease (COPD), asthma, or depressive disorders (James et al., 2018). CKD diagnoses resulted in 1.2 million deaths in 2017, a number that has been projected to rise by 2040 to 2.2 million in a best-case scenario and up to 4.0 million in a worst-case scenario (Foreman et al., 2018). Persons with end-stage renal disease are exposed to multiple physical and psychosocial stressors. Treatment in the form of dialysis poses considerable stress, including potential changes in family relations, social interactions, and occupational demands (Untas et al., 2011). Depression, anxiety, suicide, and delirium are common complications observed in patients with renal failure (De Sousa, 2008). Services of a physician and pharmacological interventions are inadequate to deal with these psychosocial issues.

This study was proposed to help patients with end-stage renal disease (ESRD), receiving treatment in a public hospital in India, deal with their psychosocial needs. Illness perception is predictive of psychosocial and clinical outcomes in patients with ESRD, including depression, nonadherence, and even survival (Chilcot et al., 2011). By applying cognitive therapy principles, the planned psychosocial intervention was intended to change maladaptive illness thoughts into adaptive ones. The intervention was developed through consultation with experts (Fraser & Galinsky, 2010) and was delivered by a trained clinical social worker when study participants attended the hospital for dialysis. Outcomes related to the extent of mental health concerns were assessed with tools measuring quality of life, emotional distress, and illness perception. The interdisciplinary clinical team members were supportive but not active participants in the intervention.

Statement of the Issue

In this research context, the intervention was designed to be implemented on the days of the study participant's visit to the hospital. During the pilot study period, the research team found participants had an unwelcoming attitude towards the psychosocial intervention. Along with the illness, many of them had significant financial constraints. Their distress, along with their feelings of hopelessness and helplessness, were strong and gave them a negative belief about a supportive intervention. On some occasions, inconvenience and lack of willingness of caregivers were a hindrance to offer the intervention. The researcher also was concerned about the randomized clinical trial (RCT) nature of the research as the persons in the control group would not be provided the intervention.

Outcomes

The research protocol was modified to address the issue of negative attitude and the disinterest of the participants by incorporating more rapport-building sessions with the persons with ESRD and their caregivers. The research protocol was ratified by the ethical committee of the hospital and that of the research center. Once the study protocol was modified, new study logistics were implemented. The study participants were recruited based on the referral of the treating physician with a clear description of the potential benefits of the intervention. All the persons referred by the treating physician were informed about the study purpose, potential benefits, risks, and their rights to withdraw from the study at any point of time during the process. Informed consent was obtained from both the participants and significant others in the family. Random allocations were made to the intervention and control groups. A convenient time for the intervention was chosen based on the preference of the participants. Intervention delivery was deferred whenever the physical distress of the participants was significantly high based on their self-report. The study considered the participants in the control group as a waitlisted group, which would be entitled to receive the same intervention as the experimental group upon analysis that demonstrated a benefit from the intervention.

Questions to Consider

1. How can participant-related concerns, such as negative attitudes about an intervention, be better addressed in the research protocol?
2. What are the necessary steps to address ethical issues emerging in a clinical research context, such as participant reluctance during recruitment?
3. Should an LAR be used in this study or regular consent? Why?

4. How do comorbidities impact decision-making for potential participants? What should a researcher consider in this context?
5. Consider how equipoise—the ethical basis for assigning people to different interventions, based on researchers' belief that, at present, no intervention is known to be better—was addressed in this study. How might you address concerns about equipoise in a study within your research area?

Case Study Analysis: A Perspective from the United Kingdom by Peter West-Oram, Caroline Brooks, and Valerie Jenkins

Context

Our commentary is based on our collective experience as researchers and a research ethics and governance professional currently serving on a U.K. medical school research ethics committee. Our response is written from a position which assumes having received the project for review at our research ethics committee *prior* to the pilot's commencement. We explain the reasoning underpinning our response and conclude that the study would not be granted ethics approval in the original application since it does not accord sufficient regard for the welfare of its participants and is likely to induce unnecessary harm. As originally proposed, recruitment was challenging, and thus the findings could be considered to have limited scientific value because they may apply to only a limited subset of patients with ESRD who were willing to enroll. It should be noted that retrospective ethics approval is not permissible and were the researchers employed at an academic institution in the U.K., a request for retrospective ethics review and approval of a piloted research study would be deemed research misconduct.

Country-Specific Regulations

Hypothetically, if it were conducted in the U.K., the study would be required to undergo rigorous scrutiny undertaken by several committees, each assessing the project and conferring differing forms of approval: scientific, governance, institutional, and ethical. As part of this review process, factors such as the merits of the research, its likely outcomes and value, the identity of participants and their potential vulnerabilities, risk factors and whether the project was generally safe and non-exploitative would be considered. While research is an important public good, it must not take precedence over the well-being of participants, researchers, or wider social values.

Addressing Ethical Issues Involving Participants in Research

As presented, this case study raises significant concerns, which would likely result in a U.K. medical school research ethics committee's decision to not approve the research without substantial revision. First, we consider end-stage renal disease (ESRD) patients to be extremely vulnerable participants. Given that the study focuses on a mental health intervention in a group acknowledged to exhibit a range of mental health vulnerabilities, participation may be likely to cause distress in both patients and their caregivers. Patients also appear to have lacked clear information about the study intervention, which no doubt contributed to their reluctance to participate. These factors suggest that the researchers had not actively trialed the feasibility of the intervention or the participant-facing materials by consulting representatives of target populations prior to the onset of the study. This lack of pilot testing led to potentially avoidable flaws in the protocol design, such as a lack of consideration for the specific vulnerabilities and needs of their target demographic. To illustrate, the intervention was originally intended for delivery on the same day that participants received dialysis. While this approach may have been designed for participant convenience—with recruitment and enrollment occurring while at the hospital for a clinical appointment—on these days, patients are likely to feel significantly more ill (and emotionally vulnerable) before receiving clinical treatment. For participation in research to coincide on these days may, therefore, be especially burdensome and psychologically distressing. As acknowledged in the "Statement of the Issue," this may have led to participants and their caregivers responding less enthusiastically to the intervention, potentially affecting results, and undermining the validity of the study.

The design of the pilot study does not appear to have adequately considered the demographic profile of the participants, their vulnerabilities, an appropriate method of approach and recruitment, as well as the potential for psychological distress and how this could be prevented. The recruitment and enrollment protocol depends on the existing relationship with the patient, but relegates participants to an entirely passive role, to be observed at the convenience of the researchers, rather than as active co-producers of research data (West-Oram et al., 2020), and more importantly as people with significant health needs and vulnerabilities, for whom participation carried significant costs, particularly in terms of psychological distress. Fundamentally, safeguarding participants' well-being should be a primary concern during study design, and the researchers' early experience highlighted several ways in which the protocol needed to be adapted to better address participant well-being. After the revision of the protocol, the researchers' attention to the ethical obligation regarding a duty of care led to more substantive engagement with potential participants to ensure a fuller understanding of the study.

For our hypothetical ethics committee to approve the study, we would require the researchers to substantially redesign their protocol and methodology to prioritize participants' needs and well-being. To achieve this, we would recommend that they engage in patient and public involvement (PPI) at an early study design stage (Cochrane Training, 2019) and work collaboratively with the public, specifically with relevant stakeholder groups, including patients, caregivers, and clinical staff (Fleissig et al., 2006). This process is undertaken to achieve five outcomes:

1. acknowledge the vital contribution of these stakeholders;
2. establish and maintain clear communication with potential participants throughout the recruitment and research process (Catt et al., 2005);

3. ensure the study does not impede the delivery of other important care, is non-exploitative and empowers participants as co-producers of research data;
4. ensure that the study intervention was delivered at a time which acknowledged participant needs (likely on a day when they have not undergone dialysis treatment);
5. ensure that the proposed intervention is perceived as potentially useful by participants, their caregivers, and the clinical team.

The study aims at an admirable objective, but was undermined by an initial design that did not adequately consider the social and psychological impacts of the research on its vulnerable participants. The PPI process we proposed, if implemented at the design stage, might have avoided the early enrollment challenges encountered in the pilot. The PPI process can be used in study design to demonstrate both feasibility of the protocol and respect for participants by considering their input as the protocol is developed. As noted above, a U.K. ethics board would require the pilot to undergo review before enrollment started and would likely not approve the original design without documentation of steps that demonstrate respect for participants—especially protections related to the social and psychological impacts—that were included in the modified (post-pilot) protocol.

Case Study #2 (Turkey): Occupational Health and the Responsibility of Researchers by Abdullah Yildiz and Berna Arda

Context

It is believed, especially in recent years, that the mental aspect of employee health is extremely important and due attention should be paid to it. In this context, a group of researchers planned a survey to assess the mental health of employees, assessing outcomes of sleep problems, anxiety, and depression. Planned evaluations included assessment of the impact of different business lines and shiftwork on workers' mental health. A metropolitan city with organized industrial zones was selected for study. Ethical committee approval was obtained for the study, as well as permission of the administrators in the zones where the study would be conducted.

While evaluating the research findings, the scores related to sleep problems, depression and anxiety were found to be significantly higher in individuals working in shifts, in accordance with the hypotheses. However, some individuals had scores high enough to warrant concern regarding serious depression and suicide risk. Of particular concern to the researchers was the fact that some of these individuals were armed security staff.

In light of this situation, the researchers sought to contact the individuals with higher scores and provide them with guidance on seeking care. However, when they were contacted, these individuals seemed overly agitated and wanted to leave the study, saying that they did not have any problems. On further investigation, the researchers found that these individuals were concerned that their employment contracts would be terminated, or they would have difficulty finding future employment if they were to undergo a psychiatric assessment. The researchers were faced with a serious dilemma in this context.

Statement of the Issue

In this study, it was understood that some information (detailing assistance and referral to support, if necessary) was not clearly provided when the participants were informed about the study. However, it can also be foreseen that such information might prevent some high-risk individuals from participating, thus excluding them from the study and possibly from referral to assistance. Further, loss of enrolled individuals who are concerned about the employment impacts they might experience related to testing for, and diagnosis of, a mental health concern would lead to loss of important data for the study.

Outcomes

It became clear that comprehensive information should be provided during the informed consent process to the participants regarding the possibility of having a screening score of concern and potential next steps. In the work setting, it is common for employees to conceal their mental problems (Huston & Larsen, 2014). To overcome this problem, it may be crucial to start with the study taking into consideration the national occupational health and safety legislation. For example, information regarding the study can be provided by contacting the occupational physicians in the relevant zone. As per the applicable legislation, it is necessary for the occupational physicians to conduct studies and guide employers regarding psychosocial problems in their areas (Legislation Information System, 2013). In this context, such cooperation will help to eliminate the potential victimization of employees and facilitate the resolution of potential problems by sharing aggregated study findings, as well as applicable responses that can be deployed, directly with the relevant businesses. Additionally, those involved in the study and charged with mitigation of occupational health issues will need to continue educational efforts to address stigma and negative perceptions regarding mental health problems.

Questions to Consider

1. Should the individuals with scores warranting referral for mental health care be removed from the study or should other solutions be sought? What might these other solutions include?
2. If it were you, how would you plan this study?
3. What opportunities exist for engaging stakeholders in the study design phase to overcome problems related to stigma, referral to care, or concerns about negative impacts on employment?

Case Study Analysis: A Perspective from India by Harneet Kaur Dhillon

Context

The mental health of employees is a neglected issue in many industries. In this study, a survey was conducted to assess mental health, specifically, anxiety, depression, and sleep issues. This survey was distributed in an industrial section of the city to study the way different business lines and shiftwork affected employees' mental health. As hypothesized, the results showed that mental health issues were more prevalent in those doing shiftwork. However, some individuals' scores were concerning for an elevated suicide risk. Some of these employees were armed security staff. The researchers attempted to contact these individuals to direct them to proper resources, but the individuals were angry and no longer wished to participate in the study. The participants were concerned that psychiatric assessment could lead to potential loss of employment and difficulty with finding jobs in the future.

Prior to the beginning of the study, the possibility of finding employees with serious mental health concerns that would require care and the need to de-identify these individuals to direct them to care was not considered carefully. Participants were not informed of such a scenario. Researchers were concerned that such disclosures could lead to a decrease in participation, particularly of individuals with more serious mental health issues, thus compromising the results of the study and allowing those individuals to continue work and life without the care they need.

Informed consent is an important pillar of protecting research participants from studies that could cause harm to them physically or psychologically. Participants should have been informed of potential findings from their responses and allowed to decide whether they wanted to participate. This process could lead to self-selection bias and the researchers need to weigh the impact of bias as they consider how to adequately protect participant privacy (i.e., disclosing study participation) and confidentiality (i.e., releasing private or personal study information). There is reasonable concern that such individuals would never receive care if they did not participate in the study; however, this issue raises the question of autonomy. Participants are allowed to make decisions regarding their own lives even if they do not seem sound to others if those individuals are deemed to have capacity which, presumably, these participants did. There are multiple ethical concerns raised in the design of this study that should be addressed with a proper informed consent process. Anticipated results of the intervention or data collection should be thought of prior to beginning the study to let participants know what to expect and how they will be informed of their personal results, if appropriate.

Country-Specific Regulations

The Indian Council of Social Science Research (ICSSR) in India was formed in 1969. Its aims were to promote social science research in the country, not to regulate it. There are no official guidelines or requirements for conducting social sciences research in India at the national level. The ICSSR is willing to provide training on how to ethically conduct research and also provides grant funding for social science research (ICCSR, n.d.). There are requirements for clinical trials, but since this type of study would be excluded from that review, it is unlikely that this study design would ever be seen by a review board prior to implementation unless wished by the scientists themselves. The institutions sponsoring this research may have some requirements in place, but if there is no regulating body that has power over them, it is difficult to evaluate whether the rules the institutions have put in place are adequate. If research is sponsored by the private sector, then it may be even more difficult to assess protection of participants. Some journals may require an adequate review process of study design prior to publication; again, if the researcher is not interested in publication in those specific journals, there is no clear incentive to design studies with ethics in mind. This issue is highly concerning, since mental health is as significant as physical health. The lack of laws and regulations surrounding social sciences research in India may undermine the significance of the social sciences. Institutions conducting social sciences research should design their own guidelines for researchers to follow, have potential studies be presented to a review board, and continue to work to have a national agency that can regulate social sciences research in the public and private sector (Srivastava, 2020).

Addressing Ethical Issues Involving Research Participants

Informed Consent

Informed consent is often obtained after having a conversation with a participant to ensure they understand the nature of the study. This process needs to be documented. In India, the National Statistical Office found that the literacy rate in 2017–2018 was around 77 percent with males being closer to 85 percent and females at 70 percent (TOI, 2020). This low literacy rate poses a challenge to gaining informed consent using written materials. If the participants are unable to read research materials provided, they are not able to give consent. The gap between males and females is concerning since females are less literate, and it is similar to the gap between rural and urban residents, with rural residents being less literate. If these groups are part of the participant pool, then additional considerations on how informed consent should be obtained are necessary. Because of the low literacy rate, consent is often obtained verbally. There are hundreds of languages and dialects being spoken in India and finding translators can be a serious concern (Srivastava, 2020). Even if the researcher is attempting to properly obtain consent, it is possible that not all the risks are being translated properly or discussed adequately. Without national regulations, it is challenging for researchers to know what expectations a review board will consider and assess whether their protocol meets those expectations; more importantly, lack of standardization of regulations can put participants at risk (Thorat & Verma, 2017). In this scenario, researchers in India need to have conversations with participants regarding the study and, if needed, find a translator from the community who can speak to participants directly.

Cultural Competency

Mental health is stigmatized in most countries, including in India (Srivastava et al., 2016). Admitting the presence of mental health conditions such as depression and anxiety is difficult for most individuals. There are reasonable concerns surrounding how one will be seen by their family, friends, employers, and society at large. Mental health diseases are seen as a consequence of lack of self-discipline or other internal causes as opposed to a result of the interactions between nature and nurture. People may not be willing to admit that there is a problem regarding their mental health for fear of retribution or ostracization. They may not recognize the signs of depression and anxiety as they have been defined by the Western world. The survey questions need to be phrased in a way that aligns with the understanding of the Indian people regarding mental health. The potential results of the surveys also need to be discussed prior to distributing them so participants have a reasonable understanding of what to expect. Resources that could be provided if participants' responses are concerning for self-harm need to be identified and should be ones that can be easily accessed by participants, if desired. Recognizing that these stigmas are entrenched in Indian society is crucial as it will likely affect participants' willingness to take part in the study and their responses, thus appropriate measures should be taken to combat such ideas and biases.

Case Study #3 (United States): Rural Organizational Research: Confidentiality and Minimizing Risk among Vulnerable Populations by Bobby J. Younce

Context

A clinical social worker was hired by a community-based mental health organization to conduct longitudinal research over three years, assessing organizational culture and effectiveness of the organization's community treatment team. The community treatment team was an interdisciplinary team of medical and mental health professionals who worked primarily with individuals experiencing severe and persistent mental illness (i.e., psychotic disorders, schizophrenia, bipolar disorder). Community treatment teams aimed to reduce hospitalizations and support independent living while keeping clients in the community. Community treatment team clients being served lived in a conservative, rural mountain town with minimal resources, no public transportation, and generational poverty. Most community treatment team clients lived with family members due to financial strain. Client satisfaction surveys were distributed to clients followed by an in-person interview with the researcher to identify service needs and effectiveness. During in-person interviews, several clients disclosed identifying as lesbian, gay, bisexual, or transgender (LGBT). Clients identifying as LGBT reported feeling isolated and stigmatized due to their sexual and gender minority (SGM) status. One client, who disclosed being transgender, was recently hospitalized after self-mutilating their genitals. Health-care providers and employers were legally allowed to refuse services or fire LGBT individuals based on their SGM status in this state. Clients asked the researcher not to disclose their

SGM minority status to the community treatment team staff or their family members due to fear of discrimination. In addition, clients feared disclosing their SGM status would result in termination of services, homelessness, or institutionalization due to a lack of acceptance from family and staff. After conducting one-on-one interviews with community treatment team staff, it became clear the organizational culture was anti-LGBT. One staff member minimized the aforementioned transgender client's hospitalization stating the client was not transgender and was merely delusional. At the time, no LGBT services were available in this community.

Statement of the Issue

In this research scenario, community treatment team clients receiving mental health services disclosed SGM status and a need for services to the researcher. It is well established that identity concealment among LGBT individuals is a significant barrier to accessing health care, while transphobia increases depression, anxiety, and suicidality among transgender persons (Budge et al., 2010; Chope & Strom, 2008; Kidd et al., 2011; Sánchez & Vilain, 2009). Given the anti-LGBT climate of the organization, lack of legal protections, and possible loss of mental health services associated with disclosure of SGM identity, the researcher must assess the risk associated with maintaining confidentiality or risk breaking confidentiality by disclosing the need for LGBT resources to the organization serving these vulnerable populations.

Outcomes

The research design was adjusted to include organization-wide interviews with staff members. Staff diversity trainings were recommended with a focus on intersectionality (i.e., multiple types of disadvantage) while maintaining confidentiality of participants interviewed from the community treatment team. The researcher also recommended the organization begin offering an LGBT group due to a lack of LGBT resources in the community.

Questions to Consider

1. How would this type of ethical dilemma be handled in your research setting?
2. What specific steps would your research team take to minimize risk of harm and ensure confidentiality among this vulnerable population while conducting organizational research?
3. How would you develop a list of resources to provide to participants?

Case Study Analysis: A Perspective from Hungary by Courtney Rada

Context

The struggles outlined in the case study all apply when considering the case in the context of Hungary. In fact, the struggles may actually be further exacerbated, as Hungary has fewer legislative protections in place for lesbian, gay, bisexual, transgender, and queer (LGBTQ+) populations when compared to many other Western nations. There have been violent attacks against LGBTQ+ identifying people in Hungary (Hungarian LGBT Alliance et al., 2015; Renkin, 2009). Due to a long history of state-sponsored discrimination and stigma against LGBTQ+ populations, Hungarians are often hesitant to reveal their LGBTQ+ status, especially to a researcher they do not know or trust. This inclination to conceal their LGBTQ+ identity often results in mental health issues and an inability to properly treat them, similar to what is discussed in the case study. Furthermore, when people choose to reveal their LGBTQ+ status, they do so at the risk of health-care providers, friends, and family mislabeling them as delusional or mentally ill. The prominent LGBTQ+ stigma, combined with relatively little legislative protection for LGBTQ+ people, makes disclosing LGBTQ+ status in Hungary a risky endeavor.

Country-Specific Regulations

In Hungary, no regulations exist that specifically provide guidance for inclusion and protection of LGBTQ+ people in research. In 2007, a health reform bill passed that resulted in closure of the Hungarian National Institute of Psychiatry and Neurology, the country's largest psychiatric teaching, treatment, and research institute, resulting in a 25 percent reduction of psychiatric hospital beds. Downstream impacts included a workforce shortage for mental health services, and a pause in progress for mental health research and treatment access (Stubnya et al., 2010).

The current social environment in Hungary makes disclosure of LGBTQ+ status difficult due to longstanding anti-LGBT legislation that criminalized LGBTQ+ behavior until 1961 (Miklóstitle, 2006), as well as recent legislation that continues to stigmatize and discriminate against LGBTQ+ people by banning their ability to adopt children, ending legal recognition for gender changes, and banning them from appearing in school educational materials, television shows, or advertising for children under 18 years (Rankin, 2021; Transvanilla, 2020). In such an environment, disclosure of LGBTQ+ status during research could well have impacts beyond the research setting. This situation makes maintaining the privacy of participants imperative, and researchers may look to general guidance on the protection of sensitive information, particularly in settings where multiple data sources may be linked, such as when research records are stored in the same system with medical/health records (CIOMS, 2016).

Addressing Ethical Issues Involving Research Participants

Using the Belmont Report framework of beneficence, autonomy, and justice can help in determining how to approach the challenge of providing mental health services to LGBTQ+ individuals living in an environment of dual stigma (i.e., mental health needs and LGBTQ+ status). As noted in the case study, a primary concern related to the principle of *beneficence* is to provide appropriate mental health services when these needs are identified in the course of the research. Given the social context surrounding LGBTQ+ identification, it is conceivable that both patients/clients and medical/service providers may not appropriately identify mental health concerns that could lead to inappropriate care delivery that could actually worsen the health of LGBTQ+ patients (Lucksted, 2004). In the Hungarian context, medical students with better knowledge of LGBTQ+ issues, as well as those with close LGBTQ+ acquaintances, were associated with less negative attitudes about LGBTQ+ individuals (Szel et al., 2020). These authors suggest that because attitude drives behavior, it is important for trainees to (1) gain knowledge about LGBTQ+ health issues to shape empathy for LGBTQ+ patients, and (2) have role models whose practice is inclusive and affirming for LGBTQ+ patients and clients to guide their own professional development. For the situation discussed in the case study, this could be achieved by creating an interprofessional team that has appropriate knowledge about, and affirming attitudes for, delivering mental health care for LGBTQ+ individuals.

Accessibility and discretion are both required for LGBTQ+ patients to feel safe in seeking out mental health care. Numerous reports have documented an ongoing shortage in Hungary of health-care professionals in general and in mental health care specifically (OECD, 2019; Svensson et al., 2017), despite a 2018 commitment that addressing mental health is one of five priority areas for the National Health Programme (OECD, 2019). Additionally, many people report that it is difficult to find providers who specialize in LGBTQ+ issues in Hungary and to maintain discretion regarding their LGBTQ+ status while receiving services. These barriers make it difficult for LGBTQ+ populations to get the proper therapy and psychiatric attention they require. This lack makes navigating their SGM status even more difficult, further exacerbating many of the difficulties these populations face (Kidd et al., 2016).

The ability to choose when to access mental health services and to control whether LGBTQ+ status is disclosed during treatment is related to the principle of *autonomy*. Both the research and mental health teams will need to respect participants' decisions about whether and when to access mental health services. Further, these teams will need to provide an environment in which sensitive data are adequately protected (CIOMS, 2016). Research data should be stored separately from other data, such as electronic medical records, whenever possible to ensure that disclosure of LGBTQ+ status is controlled by the participant and accidental disclosure is avoided. Research findings should be reported in aggregate, and data sharing for additional research should be carefully controlled by using individual participant de-identification and maintaining records of what data are shared and with whom. These protections should be clearly explained to participants during the informed consent process, along with the researchers' plan for analyzing and presenting the findings. Providing research participants with some autonomy in the way the research is conducted may also be beneficial. For example, researchers may allow the participant to choose when and where to participate in an interview and allow them to opt out at any time. Demonstrating adherence to these principles builds trust between researchers and participants, creating a culture of safety for future participants who have expressed concerns that research participation exposes them to additional risk of stigma.

As highlighted in the case study, LGBTQ+ individuals are at increased risk of poor health and social outcomes when they are not able to safely access needed mental health services. The principle of *justice* asserts that benefits and burdens of research participation should be distributed fairly. When they do not participate in research that is potentially beneficial to them because of fear of socially detrimental outcomes, LGBTQ+ individuals are being disadvantaged. Thus, it is important for researchers to advocate for conditions that support LGBTQ+ individuals to participate in research. In Hungary, as in other countries, a three-pronged approach to ethics is needed that will address research, clinical practice, and ongoing training (McGinley & Horne, 2020). Specifically, research will advance as (1) the social environment improves for those identifying as LGBTQ+, and (2) the clinical environment is better able to address LGBTQ+ health through workforce awareness and skill (Szel et al., 2020) and adoption of best practices in mental health policy and health care (Rodzinka, 2020).

Summary of Chapter

This chapter highlights a variety of issues to consider when conducting research on mental health or people who are experiencing mental health issues. Research outcomes that result in diagnoses of, or indicators of concern for, mental health conditions require a research team to explain the potential for these outcomes to people considering enrollment. Thinking through each scenario, how it may positively or negatively impact a participant, and developing resources for participants that can help them navigate the diagnosis/finding are important steps in protocol development, preparation for ethical review, and study implementation. This level of detail may be reassuring to some and concerning to other participants, but providing this detail is essential for potential participants to evaluate the risks and benefits of study participation and provides an opportunity for discussion that supports informed consent. Research on mental health conditions—stigmatizing on their own—in combination with other individual identities that are linked to discrimination or stigma can further increase participants' vulnerability in the research setting. In situations where potential participants may be viewed as vulnerable, extra steps should be taken to implement safeguards to protect the rights and welfare of these individuals. These steps may include thoroughly explaining research activities and potential outcomes; providing additional time for considering study enrollment; conducting research activities in safe spaces; creating and explaining safeguards for participant privacy and protocols for confidential storage and use of data; and honoring participants' decisions to opt out of some or all research procedures at a later date.

The case study from India re-evaluated informed consent and enrollment activities, in recognition that these did not initially fully explore the impacts of study participation for a group of patients with end-stage renal disease. The potential benefits of participation needed to be balanced with the added burdens of participation. These burdens and benefits needed to be conveyed to both participants and their caregiver/family support people. As noted in the case analysis from the U.K., much of this work of envisioning potential outcomes and impacts can be accomplished even before pilot work begins by engaging with similar types of patients/participants and stakeholders from groups that work with them in the domains of clinical and social service delivery. The information gained in this process can guide the informed consent process and development of patient-facing materials for both recruitment and navigating changes in health or social service needs throughout the study.

The second and third case studies touch on the need to protect privacy (of individuals) and confidentiality (of data) in mental health research. In the case study from Turkey, questionnaire-based research resulted in the identification of individuals who may be experiencing depression or anxiety. However, because of stigma associated with a mental health diagnosis and the concern that such a diagnosis might negatively impact employment, participants were resistant to referral for a more thorough evaluation. In the case study from the U.S., participants involved in a community-based mental health intervention were afraid to disclose their LGBTQ+ status over concerns that they would be rejected by people in their social network and would also not receive appropriate care in health-care settings that were hostile to, or at a minimum unfamiliar with, LGBTQ+ appropriate care. Both analyses (from India and Hungary) reiterated concerns about stigma and negative impacts from disclosure of mental health or other stigmatized conditions, especially if this disclosure was not controlled by the participants. These analyses emphasize the need for explaining how study data are stored, used, and safeguarded, mentioning who has access and how the data will be reported.

Overall, these case studies and analyses highlight the importance of setting expectations for potential outcomes in studies that collect sensitive information, and adequately framing the risk–benefit analysis for potential participants' consideration, including discussion of participant privacy and data confidentiality. Further, the variation among countries regarding the ethical review processes demonstrates that expectations for review—and standards for that review—can be quite different. Researchers should, therefore, be prepared to assess the local situation with regard to ethical review processes and sociopolitical contexts surrounding data collection of information that may be deemed sensitive.

Chapter Exercises

1. In your own research area, identify topics that might be considered sensitive. What regulations or guidelines in your country might guide your decision-making about elements to include in an informed consent discussion with potential participants about the benefits and risks of participation?
2. How can research records be kept separate from medical or employment records in your country? If they can't be separated, what kinds of safeguards can you implement to protect the confidentiality of research data?

3. In your own research area, how might you frame a discussion with your colleagues about the reason(s)—i.e., concept of equipoise—for a comparative study about a mental health topic?
4. Discussion of intersectionality is becoming more common in research. How might participants' different identities impact how you develop research in your country with regard to an identity such as race? Consider some other identities, such as socio-economic status, sex and gender identities, and disability, or another identity related to potential participants. How do these considerations further refine how you might develop your research?

8 Genetics and Biospecimens

The collection and storage of biospecimens for the purposes of research has provided unique research opportunities and scientific advancements; however, it also presents an interesting ethical quandary about the rights people have to their own tissues once they are collected. A particularly public example of this debate unfolds in the story of Henrietta Lacks, whose tissues were unknowingly appropriated for research. These cells, known as HeLa cells, were cultured and have become one of the most widely used biological samples (Beskow, 2016). Since this incident, regulations concerning biospecimens have advanced, but there remains much debate on the subject.

There is, however, little debate as to whether informed consent should be required in order to retrieve a biospecimen from an individual. The Declaration of Helsinki requires informed consent when using biospecimens for research except in the most extreme cases (WMA, 2018). Likewise, the Council for International Organizations of Medical Sciences (CIOMS) requires informed consent when collecting biospecimens, but also allows for broad informed consent when specimens are collected to be used in some unspecified project in the future (2016). These guidelines allow for the use of an informed opt-out procedure when using residual tissues left over after medical procedures, rather than the traditional informed consent or even broad informed consent. With an opt-out procedure, consent is assumed, and participants must voice their dissent in order to prevent their specimens from being used (CIOMS, 2016).

Broad informed consents allow researchers to utilize collected specimens to investigate research questions that were not known at the time of biospecimen collection. In the U.S., the use of existing de-identified specimens, in which it is impossible to determine who the specimen came from, is exempt from informed consent, since the research is not considered to be human subjects research (OHRP, 2008). The CIOMS guidelines also allow for informed consent to be waived when using de-identified data, as long as the research promises to have important social value.

In 2017, the final version of the Common Rule in the U.S. was published. There are several new parts, including those to the informed consent pertaining to information and biospecimens: "1) A statement of whether biospecimens may be used for commercial profit and whether subjects will share in the profit, 2) A statement about whether results will be disclosed to participants, 3) A statement to inform participants if their biospecimens will undergo whole genome sequencing" (NIH

DOI: 10.4324/9781003033233-8

2017). The Final Rule also addresses the elements of broad consent for biospecimens. Under this, researchers are not able to use a participant's samples or information if they do not sign a broad consent.

Since the mapping of the human genome, the fear of misuse of genetic information has become more prevalent. The potential for the abuse of genetic information has been recognized by many countries, as evidenced by the Genetic Information Nondiscrimination Act, passed in the U.S. in 2008, as well as similar acts more recently passed in Canada (Genetic Non-Discrimination Act S.C. 2017, c.3) and the European Union (EU) (EU Charter of Fundamental Rights, Article 21: Non-discrimination). These acts prevent health insurance companies, life insurance companies, and employers from discriminating against people because of a supposed undesirable genetic trait. Because of the sensitive and contentious nature of genetic information, care must be taken to protect participants' confidentiality, as with any personal information collected during the course of research. The Common Rule also has regulations requiring participants be given information on the genetic research being done, how their samples will be used, and the full implications of their participation, both risks and benefits (NHGRI, 2017).

There are several companies that offer genetic testing and with these come some concerns when it comes to the impact of having genetic information collected by third parties. Although parties such as 23andMe and Ancestry.com have to follow the Common Rule and require anyone interested in participating to sign a full consent document, their privacy statement notes that the information can be used for other research (23andMe 2019; Ancestry, 2019). The National Institutes of Health (NIH) and an institutional review board (IRB) have the right to audit the data.

A unique aspect of genetic research that has presented a particular ethical challenge for researchers is the instance of unsolicited or incidental findings, and whether to inform participants of any type of findings. These occur when something is uncovered during the course of research which is outside of the scope of the project, but which may be of clinical significance to the participant. In studies where there is a potential for unsolicited findings, this consequence should be taken into account in the consent form. CIOMS guidelines suggest the use of a "tiered consent," which gives multiple options for participants to decide how much information, if any, they would like to receive if the research team discovers something significant (2016). If the potential for unsolicited findings is not addressed in the informed consent form, researchers must decide for themselves whether it is more unethical to re-contact participants without their consent or to allow them to remain ignorant of potentially serious health information. For example, Ancestry.com outlines in their terms and conditions about their services that although they have ownership of their data, a participant could discover unanticipated facts about themselves or their family while using their services and this is something participants cannot change (Ancestry, 2019).

According to Nilsson et al., regardless of the status of the consent form, there is general consensus that in order to justify re-contacting research participants, findings must fit three criteria (2018). First, results must be analytically valid; meaning that the chances of false positives are low. Second, results must have clinical significance; there must be meaningful evidence connecting the genetic information to the occurrence of disease. Third, there must be immediate action that can be taken in light of the information. If all of these criteria are fulfilled, it may be the ethical choice to re-contact participants to notify them of findings. If the participant who provided the genetic sample is deceased, it may be necessary to contact their family members to notify them of their risk (Nilsson et al., 2018).

In the case of research in low resource settings, the question of re-contact becomes more complex. Determining how to inform participants about what the results mean to them can be difficult, since health literacy about genetics may not be very high. In addition, results that may be considered "actionable" (still open to clinical discussion) to one population may not be as easily dealt with in another. There is debate considering the extent to which the research team is responsible for providing counseling and management.

The foci of the case studies in this chapter are:

- broad consent or multiple consents for use biospecimens and the language involved;
- genetic information abuses—data sharing; and
- the use for or against a therapy with inadequate evidence.

Case Studies and Analyses

Three countries (Jordan, Sweden, and the U.S.) are represented in the presentation of these case studies. The case studies in this chapter deal with (a) the vagaries of the potential impacts of genetic testing when populations are screened for sickle cell trait for the purposes of resource allocation, (b) high-income-country-funded projects serve to alleviate diseases spread from low- and middle-income countries (LMICs), however, the technologies applied in the high-income country may be unavailable in the LMIC where the technologies were developed, and (c) the use of biospecimens in the lab to extrapolated clinically untested therapies.

The case study analyses in this chapter come from the points of view from researchers in three countries (the U.S., Croatia, and Australia). For the case study analysis from U.S., concerns are raised about the accuracy of the consent process in a culture with a different concept of autonomy. In the Croatian case analysis, availability of vaccine technologies is demonstrably inequitable and proposed solutions such as open access research fails to guarantee partners equal benefits. Finally, in the Australian case analysis a discussion regarding the ethical issues related to equal access and health equity, standards of evidence, and resource allocation is presented.

Case Study #1 (Jordan): Ethical Issues with Informed Consent Regarding Biospecimen Collection by Qosay Al-Balas

Context

Mediterranean countries are considered, in general, as conservative and have a tribal system that regulates much of the internal operations of families. A study, conducted by a university researcher, included taking blood samples from patients referred to a hospital to determine the distribution of a hereditary disease (sickle cell anemia) that is common in this country. The aim of the study was to investigate the spread of this disease in certain tribes and give recommendations to the health authorities trying to find suitable solutions to reduce or alleviate the incidence of this disease in the future, and subsequently reduce the cost of treatment. The consent form was IRB-approved, and there was a statement that the data collected in this research would only be seen by the health authorities. In addition, the patients' samples were coded and could not be identified by anyone other than the investigators. During the first month of enrollment, very few patients were willing to participate in this study. After investigation, it was found that patient refusals were related to concerns about how the research will affect their tribes by stereotyping them that could lead to stigma, and impact whether people would refuse to marry someone from their tribe. Moreover, females were requested to consult their husbands or parents before being involved in the study.

Statement of the Issue

During the consent process, patients were only informed that the purpose of the test was to check if sickle cell anemia was present or absent. Potential participants were told that the researcher was conducting the study "for their benefit," but details about the purpose of the research were not provided. Patients were requested to make their decision quickly, sign the consent, and provide the blood sample at that visit.

Outcomes

Low participation was improved by re-training the research staff who were administering the informed consent, and a sufficient number of samples was collected from a diverse group of referred patients. However, complaints had been made by participants, and these were highlighted in the press. Tribal members felt they had been targeted and were concerned about future stigma. All samples are currently stored at the researchers' university and are ready to be analyzed.

Questions to Consider

1. How should informed consent be handled in this situation where samples are anonymized? Could this study be conducted without a consent process?
2. Based on the ethical principle of utilitarianism (where the outcome of improved health for an entire community guides decision-making related to process), how would you decide what information about the study should be disclosed to potential participants?
3. Should the IRB stop the experiment and have the collected samples destroyed?

Case Study Analysis: A Perspective from the United States by Jahn Barlow

Context

In the U.S., the foundation of ethics for research involving human subjects is built on respect for individual rights and autonomy, assessing the risks and benefits associated with the research and principles of justice (National Commission for the Protection of Human Subjects of Biomedical and Behavioral Research, 1979). The regulatory framework is further defined in the Code of Federal Regulations under 45 CFR Part 46, Protection of Human Subjects, which creates IRBs for oversight of research involving humans.

Within this philosophical and regulatory framework, the preeminence of individual rights in the social and political culture of the U.S. influences our understanding and implementation of ethical research. One of the key individual rights in a liberal democracy is the right to privacy. In the context of the genetic research study in Jordan, the individual right to privacy can be defined as the control that individuals have over information about themselves, including their genetic vulnerability to sickle cell disease. An extension of the right to privacy is the concept of confidentiality. Researchers have an ethical obligation to protect sensitive information from further disclosure as described to participants in the informed consent process.

Country-Specific Regulations

To satisfy U.S. expectations of individual privacy and confidentiality, participants need to understand the nature of the information being collected about them and how that information will be used in research. From a U.S. perspective, a serious flaw in the initial consent process of the Jordanian study was informing potential participants that "the purpose of the test was to check if sickle cell anemia was present or absent." This simplistic binary description of sickle cell genetic testing is misleading. The difference between inheriting a single sickle cell gene from only one parent (asymptomatic carrier) and inheriting two sickle cell genes, one from each parent (sickle cell disease) was apparently not communicated. This critical distinction is important to the participants' understanding of the role of genetic testing for their own health and for the health of their descendants. Without a basic understanding of the genetic tests performed, participants were placed at a substantial disadvantage in making decisions about their own privacy as well as the potential privacy concerns of their posterity.

The privacy and confidentiality issues are further complicated by an inadequate plan for communication of the research results to participants. In light of the statement that the data collected in this research will only be seen by the "health authorities," one assumes that genetic test results will not be returned to the individuals who participate in the study. Even with anonymized samples, this asymmetry of information has the potential for abuse of individual autonomy. For example, corrupt health authorities could leak information (or misinformation) to undermine the marital prospects for individuals from competing tribes or families. Women are almost certainly more vulnerable to such abuse of autonomy in societies with evidence of severe gender inequality. Although improving, Jordan still ranks 131 out of 156 countries in key measures of gender equality (World Economic Forum, 2021). Notably, women were initially asked to consult with their husbands or parents before participating in the study, an indication of intrusion on their autonomy that raises questions about whether their participation was completely voluntary. For researchers in the U.S., there would be significant ethical concerns raised about the privacy implications for the reproductive rights and autonomy of women who are already facing high levels of political, economic or social discrimination. This study does not seem designed to empower women with information, nor does the study specifically address the interests of women who may choose to participate to contribute to the health and well-being of their families and communities without adequate consideration or recognition of their own well-being.

Addressing Ethical Issues for Involving Participants in Research

In assessing the risks and benefits of the research, the lack of details about the specific goals of the study is extremely concerning. Providing "recommendations to the health authorities" could encompass a vast array of possibilities, some of which may not be in the best interest of individual participants. The benefits of the study were initially communicated only vaguely to the participants and did not address how the study would reduce or alleviate the incidence of sickle cell disease in the future. The risk of the blood draw for the genetic test was likely considered minimal risk, but there is no indication that the researchers or the IRB took into account the psychological or sociological risks that would accompany the genetic information obtained by the study. Yet, some of the participants identified these risks as potential future harms to their families and tribes and, therefore, refused to participate in the research.

Even though the study staff was retrained on procedures for obtaining informed consent, it is not clear that all of the necessary components to ensure autonomy, beneficence, and justice were in place during the consent process. The researchers should consult with their IRB, local health authorities, and tribal leaders about reports of the tribal members' feelings that they were targeted and placed at risk for stigmatization. With the emphasis on individual rights and autonomy in the U.S., the researchers may need to consider re-consent before analyzing the samples. At a minimum, they should develop a plan to address the more egregious ethical errors in the conduct of this research. First, they should discuss whether the study and/or local health authorities have resources to provide the participants with (a) clinical genetic tests that would allow for the reliable and ethical return of results to the individual participants and (b) genetic counseling sessions with trained professionals who have developed cultural competence through training or experience to help individual participants process their genetic results and make personal decisions about how to use the information. Ideally, genetic counseling would allow for follow-up as needed to address specific concerns or risks to participants that may arise after learning their results.

If the resources for clinical genetic testing and counseling are currently unavailable, there should be an evaluation of the availability of these services to study participants in the future. If these services are unavailable or unaffordable for the foreseeable future, then the researchers and local health authorities should consider developing a plan to communicate the aggregated results of the study to the population in a way that not only protects individual privacy, but also protects the identities of specific tribes or families who could be stigmatized by increased incidence of sickle cell disease or sickle cell carriers among their members. Engaging the study participants and the broader community in dialogue will help facilitate understanding with the goal to minimize stigmatization.

Re-contacting all participants in the study may be difficult or impossible, but given the concerns raised by the tribal members, a reasonable effort should be made to contact as many participants as possible to provide adequate information about the nature of genetic testing and the specific goals of the study that were omitted from the initial consent process. Everyone who is contacted should be given the option to withdraw from the study and have their samples destroyed to respect their autonomy and promote justice to the extent possible. The IRB would be wise to pause the analysis of the samples until the researchers submit a suitable action plan with adequate input from the local health authorities and local communities who may be affected by the outcomes of the research.

At this point, it may be impossible to mitigate all of the potential issues that will arise from the mistakes made by the research team. With a sincere and concerted effort to address fundamental issues of privacy, autonomy, and justice through engagement of the local health authorities and the local communities, under the oversight of the researchers' IRB(s), the researchers have an opportunity to correct course and demonstrate the vital importance of ethical research to their colleagues in the U.S. and Jordan.

Case Study #2 (Sweden): Emerging Diseases: Sustainable Capacity Building Beyond the Scope of a Single Project by Tomas Klingström

Context

Lumpy skin disease (LSD) is an infectious disease transmitted to livestock by insect bites. It was first seen in Zambia in 1929 and spread throughout Africa in the subsequent decades, causing large economic losses. In Africa the disease is now endemic and apart from cattle, sheep, and goats the LSD virus can also be detected in wild animals such as water buffalo and impalas. Mortality rates in domestic African breeds are reported to be 1–3 percent and infection is accompanied by severe production losses. It is also known that *Bos taurus* cattle such as the Holstein-Friesian, which dominate the European dairy industry, are far more susceptible to disease with mortality rates as high as 40 percent being reported in some herds.

In 1989, LSD was detected outside of Africa for the first time in an outbreak in southwestern Israel, where 14 out of 17 dairy herds in the region were infected. An eradication effort with extensive culling in the infected region put an end to the outbreak but the disease has spread across the Middle East, and in 2012 the first cases were reported within the European Union (EU). Therefore, several European organizations and national authorities have been mobilized to deal with LSD as an emerging threat to European livestock. As a part of the European effort, the Defend2020 (2022) research project was also funded through the Horizon 2020 EU Research and Innovation program to deal with the dual emerging threats of LSD and African swine fever spreading into Europe.

Horizon 2020 research projects are multinational efforts assembling academic researchers and other stakeholders to achieve a number of objectives related to research and innovation during a limited time frame (three to five years). Defend2020 is funded from 2018 to 2023 and aims to control the expansion of LSD and African swine fever in Europe and neighboring countries. African swine fever is currently spreading throughout the boar population in Europe, but the spread of LSD has been stopped completely in Europe through the mass vaccination of millions of cattle in southeast Europe using a live attenuated vaccine developed in South Africa. In 2019, there were no new outbreaks reported in Europe. With a verified disease prevention strategy in place, the threat of LSD to the livelihoods of European farmers is greatly diminished.

Statement of the Issue

LSD will remain a significant issue in many low- or middle-income countries for the foreseeable future (Klingström et al., 2016). At the same time, much funding within the project has been invested in research that could only be carried out in a select few institutions located in high-income countries far away from the current extent of the disease.

Thus, it is important to plan for what happens after a project ends and how to ensure that the benefits of a project can contribute to both project partners and external stakeholders. This issue is especially important in projects such as Defend2020 where important project partners are both socially and geographically distant from the regions that have the most potential to benefit from the project. Data sharing and open data have, since 2017, been recognized as the norm for EU-funded projects but sharing of results does not necessarily guarantee that all partners share the benefits of a project. Any successful project must recognize these issues and capacity building should be a part of the planning process to ensure that project outputs are broadly implemented.

Outcomes

The Defend2020 project is taking a multipronged approach to disease management dealing with risk analysis, phylogenetics, epidemiology, host response, and vaccine development. Stakeholders have different interests and capabilities to integrate the results from these actions within their work. A survey is being conducted to identify key priorities among partners and training is being planned accordingly. Some partners have also decided to use the Galaxy platform for bioinformatics to code their data analysis workflows in a unified format making them more accessible to stakeholders lacking access to trained bioinformaticians to work with the data (Afgan et al., 2016).

Questions to Consider

1. Many low- or middle-income countries have restricted exports of samples containing genetic material out of fear of being exploited, as samples are often analyzed with little benefit to the country providing the samples. When should a country safeguard its genetic resources? When is it acceptable that samples are exported for analysis abroad? What should be the expected return from sharing samples in a complex project with a multipronged approach like Defend2020?
2. What is the difference between sharing the results of research and sharing of benefits from research? Under which circumstances does one not necessarily lead to the other?
3. How can projects be planned and conducted in a manner that ensures that capacity is built among stakeholders with a long-term commitment to an issue?
4. What responsibility as a researcher do you have to ensure that data benefit all stakeholders providing resources to a project?

Case Study Analysis: A Perspective from Croatia by Joanna Gorka

Context

Lumpy skin disease (LSD) is a viral disease affecting cattle and other livestock. Endemic to Africa since 1929 with movement to the Middle East and Europe since 1989 and 2012 respectively, this virus has caused significant economic turmoil. The LSD virus is transmitted by blood-feeding insects and causes skin-nodules, fever, and potentially death. Even surviving cattle are disadvantageous due to reduced milk yield, damaged hide, infertility, and incidence of spontaneous abortion. (Gibbs, 2021). The standard cattle species in the European dairy industry, *Bos taurus*, are more susceptible to LSD virus and exhibit higher mortality rates.

Efforts to control the virus include import restrictions, surveillance measures, and vaccination programs (Calistri et al., 2020). Despite success in many European countries, low-income countries lack the resources to manage their livestock populations. It is in these low-income countries that LSD is a significant issue and economic pit. The nature of viral disease in an industry that features global imports and exports necessitates a collective effort from countries. To what extent should cooperation be employed?

Tomas Klingström is a Swedish scientist involved in Defend2020, a European Union project funded from 2018 to 2023 that targets the spread of LSD and African swine fever. Klingström works with the Swedish University of Agricultural Sciences to design and facilitate open software and collaboration on genetic research in low- and middle-income countries (LMICs). In the case study "Emerging Diseases: Sustainable Capacity Building Beyond the Scope of a Single Project," Klingström argues for planning for continued action after project funding ends in 2023, as well as ensuring that countries outside the EU can benefit from LSD containment measures. This sentiment is especially relevant to Defend2020 because significant project partners are both "socially and geographically distant from the regions that have the most potential to benefit from the project" (Klingström case study above).

To identify solutions, it is helpful to reflect on examples of success. Croatia, a member of the EU since 2013, carried out an LSD surveillance program from 2016 to 2018. Croatia's economy is highly dependent on Europe's principal economies and major trade partners with Croatia including Italy, Germany, and Slovenia (Dag Hammarskjöld University College of International Relations and Diplomacy in Zagreb, 2016). Croatia's population is 4.3 million with more than half of individuals living in rural areas. Croatia's identity as a relatively small and under-resourced EU nation indicates that their successful LSD vaccination/surveillance program can serve as a model of LSD efforts that could potentially be accomplished in low-income countries.

Country-Specific Regulations

Livestock farming is present in all regions of Croatia with the majority belonging to small family farms. Cattle farming, particularly milk and meat production, is the most important sector of livestock in Croatia. Yet, Croatia is neither self-sufficient in beef production nor in milk production. The Registry of Agricultural Holdings, developed in 2002, is a registry of Croatian farmers. The registry supervises the agricultural market and is responsible for monitoring the health of domestic animal care. The Registry of Agricultural Holdings documents all transfers of domestic animals and ensures compliance with the Veterinary and Animal Welfare Act (Croatian LSD Surveillance Programme in 2018, 2018). Local farmers are faced with monetary penalties if they fail to comply with animal labeling systems and monitoring regulations (Dag Hammarskjöld University College of International Relations and Diplomacy in Zagreb, 2016).

During 2015, the import of cattle increased by 21.147 units compared to the previous year with most of the imported cattle being from Romania (Dag Hammarskjöld University College of International Relations and Diplomacy in Zagreb, 2016). Croatia's moderate dependence on cattle imports makes it particularly vulnerable to LSD exposures. Croatia joined the Standing Group of Experts on Lumpy Skin Disease in South-East Europe which started in 2016 (World Organisation for Animal Health, 2021). This organization aims to facilitate cooperation among the countries affected by LSD and provides a forum to discuss tested and untested mitigation efforts.

In 2016, Croatia began its first preventative vaccination campaign against LSD. Despite a lack of diseased cows in Croatia, this movement was initiated due to disease presence in surrounding countries. In the first campaign, 432,145 cattle were vaccinated with a live homologous vaccine. In 2017, 468,574 cattle were vaccinated reaching a 95 percent vaccination rate. Based on data provided by the Standing Group of Experts on Lumpy Skin Disease in South-East Europe, mass vaccination emerged as the most effective control policy among countries (Porcel, 2021; World Organisation for Animal Health, 2021). In agreement with the Terrestrial Animal Health Code by the World Organization for Animal Health, Croatia developed a surveillance program (OIE, 2018). In 2018, a map was designed outlining eight Croatian counties deemed as "risk areas." Six hundred and forty unvaccinated cattle were selected randomly for monthly qPCR and ELISA testing. Deceased cattle were examined for LSD pathology; LSD was not identified in any specimens. Surveillance also included temperature measurements and skin, lymph node, mucosa, udder, and perineal examination of healthy cattle (Veterinary and Food Safety Directorate Ministry of Agriculture, 2018). The surveillance campaign deemed Croatia clear from LSD and was ceased in 2018. The decision to lift the restrictions on the trade of bovine animals from Croatia has been made by the Standing Committee on Plants, Animals, Food and Feed that then decided to lift restrictions imposed on cattle trade in Croatia (Hina, 2019).

Addressing Ethical Issues Involving Research Participants

In 2019, LSD vaccination in Israel shifted from being mandatory to voluntary. Due to unprotected cattle and circulating virus in surrounding regions, Israel experienced a LSD resurgence. Therefore, the European Food Safety Authority recommends that countries develop contingency plans and stockpile vaccines (EFSA 2020; Calistri et al., 2018). However, this may not be an option for many low-income countries. The EU's Defend2020 project may serve as a partial solution, even for countries outside of the EU, if the database maintains its open-access nature. Countries experiencing LSD exposures may be guided by protocols followed by nations of similar resource banks. EU members also receive funding for LSD surveillance programs. Despite massive LSD containment in many countries, new exposures due to livestock trade are possible, suggesting that high/middle-income countries may reap economic benefits by promoting research sharing and potentially even offering aid to other entities. Klingström warns that although data sharing has become the norm for EU-funded projects, open access research fails to guarantee partners equal benefits. These discrepancies should be addressed in plan development and should be monitored through active campaigns.

Case Study #3 (United States): Emergent Conditions for Research: COVID-19 as an Example by L. Scott Benson

Context

The COVID-19 pandemic that was first described in Wuhan, China in November 2019 provoked conflict between the political system and established medical scientific dogmas. These conflicts led to numerous examples of traditional best practices being challenged or abandoned all together. COVID-19 was caused by the SARS-CoV2 virus, a newly emergent virus, about which little was known. Initial descriptions of the illness and death caused by the virus prompted many in the medical and scientific community to feel pressured to "do something" as the outbreak grew and expanded into a global pandemic. Political decisions to close borders, enforce quarantines, and issue shelter-in-place orders were simultaneously applauded as protective and condemned as governmental overreach. While most of these political decisions were made in an effort to control the spread and impact of the virus, both the decisions and the outcomes are ripe for ethical examination. Here we will focus on the medical response to COVID-19.

Statement of the Issue

As COVID-19 spread from China to surrounding countries, then across the globe, health-care providers were making decisions without guidance on tested, proven and accepted diagnostics, therapies and preventive strategies for this novel virus. It was in this setting that unproven medical therapies, such as hydroxychlorquine (HCQ), remdesivir, and lopinavir-ritonavir (Kaletra) were used, since the perception was that doing something was better than doing nothing. Concurrently, the race to develop accurate, sensitive, and specific testing strategies and safe, effective vaccines evolved as a counterpart to therapeutics. A brief rationale for the use of each of these medications is provided below:

- With the Severe Acute Respiratory Syndrome (SARS) and Middle East Respiratory Syndrome (MERS) outbreaks, HCQ had been evaluated as a potential therapeutic option. It demonstrated *in vitro* anti-viral activity when tested with a number of different viruses. It is relatively inexpensive to produce and has been used extensively for the prevention and treatment of malaria, and is used in select patients with lupus and rheumatoid arthritis. Due to the history of *in vitro* anti-viral activity, HCQ was utilized, off label by Chinese and Italian physicians desperate to stem the mortality of the disease.
- Remdesivir, an antiviral originally developed to treat Ebola, was used in hospitals around the world because it has been effective against a broad spectrum of viruses. As remdesivir is an intravenous (IV) administered medication, its utility to the general public is limited, but its use in hospitalized patients with IV lines was reasonable.
- Kalera is a well-known HIV antiretroviral agent in the class of protease inhibitors. HIV and SARS-CoV2 are both RNA viruses so there was some reason to theorize that there might be therapeutic benefit with COVID-19. As an oral medication, Kaletra would be easy to distribute to the public, but has some significant side effects, such as diarrhea, nausea, vomiting, and headaches.

In each case, these medications were developed for other diseases and conditions; however, there were theoretical reasons to hope for potential therapeutic value.

Outcomes

Results from preliminary studies were released prior to peer review and publication and were often contradictory, with some claiming improvement in viral loads and survival, while others found no benefit. With a growing number of deaths and a spreading pandemic, positive outcomes for all three medications were reported with little mention of the contrary findings. Many physicians, scientists, politicians, and members of the public pushed for increased access and utilization of these therapies, while others called for calm, deliberative study of these medications before including them on treatment protocols. Nations and states organized and began stockpiling medications in case they worked, "just in case" there would be shortages.

Questions to Consider

1. If there is inadequate evidence for or against a therapy, but there are no proven options, how can an individual be fully informed about the risks and benefits and offer consent to treatment?
2. Do providers demonstrate beneficence by providing or withholding unproven treatments from sick patients?
3. Under emergency conditions, what standard of evidence should be used to decide on the use of an experimental treatment?
4. How does early release of information from a study without peer review impact research findings in the future and treatment options?

Case Study Analysis: A Perspective from Australia by Bridget Pratt

Context

In the Australian context, testing existing medical interventions for efficacy against COVID-19 raises ethical issues related to equal access and health equity, standards of evidence, resource allocation, and fair participant selection, both at national and global levels. Studies testing existing medicines for efficacy against COVID-19 had strong potential to compromise access for existing patients with other, often serious illnesses. In the case example, the supply of hydroxychlorquine (HCQ), remdesivir, and lopinavir-ritonavir (Kaletra) for patients with HIV, arthritis, and lupus could be compromised, first, by medicines being used in clinical trials rather than for their treatment and, second, by members of the Australian public or the government buying or stockpiling such medicines because there is a chance they could have protected them from COVID-19. This raises the questions: Is it fair for patients with existing illnesses to lose their access to medicines, which, in turn, could have severe effects on their well-being and life expectancy, in order to potentially protect others from diseases such as COVID-19? Will the burden on existing patients fall disproportionately on those who are already considered to be disadvantaged or marginalized? Where governments stockpile essential medicines, this hoarding may also create problems for patients in other countries accessing those medicines to treat HIV, arthritis, and lupus and to treat diseases such as COVID-19 (if proven effective). Less access for patients in low-income countries would serve to reinforce or worsen existing health disparities between countries, which is inconsistent with principles of global health justice.

Before undertaking studies on existing medicines like HCQ, remdesivir, or Kaletra in Australia, it is important to assess how likely it is that a shortage will result for existing patients in Australia and in other countries and how big that shortage might be. If a shortage will result, the next consideration is how severe would its implications be for existing patients' health and overall well-being. If the impact would be severe, it is ethically essential to have very strong theoretical reasons to hope for potential therapeutic value and put strategies in place to mitigate the risk of shortages occurring for existing patients both in Australia and overseas.

Barriers to existing patients' access might also be generated after studies are underway. With many powerful players calling for the early release of COVID-19 research findings, companies frequently reported early positive findings in the media before studies were even completed (Haseltine, 2020). But this practice was often used as a means to inflate stock prices and could create potentially misleading impressions of their drug's efficacy, which, in turn, could promote stockpiling and use of an ultimately ineffective medication (Haseltine, 2020). Who bears the harms and benefits of early reporting in the media is likely unevenly distributed, with companies reaping the majority of the benefits.

Country-Specific Regulations

The case raises ethical issues around standards of evidence, as positive outcomes for all three medications were reported with little mention of contrary findings of no benefit. In order for an accurate picture of the efficacy of medicines like HCQ, remdesivir, or Kaletra against COVID-19 to emerge, both positive and negative study results needed to be reported. Otherwise, evidence trends cannot be accurately assessed *across* studies. It is these across-study effect sizes that are important to determining whether such medicines should be used as diseases such as COVID-19 treatments. Ideally, the evidence trend should be towards showing significant benefit, rather than showing a drug is equally likely to offer a benefit or not. Regulators like the Therapeutic Goods Administration also need to know about negative study results to make informed decisions.

Under intense, emergency conditions like those experienced in the COVID-19 pandemic, regulators may experience severe pressure to amend existing standards of evidence to get treatments approved for use more quickly. In the case example, physicians, scientists, politicians and members of the public pushed for increased access and utilization of HCQ, remdesivir, and Kaletra. Depending on the balance of power, regulators may or may not be swayed by these pressures. The case also raises the questions: What standards of evidence for regulatory approval should exist in an epidemic or pandemic context? Are they lower than in non-emergency contexts? Are we willing to accept greater risks for less proven benefits?

Addressing Ethical Issues for Involving Participants in Research

Testing of existing medicines for efficacy against COVID-19 occurred alongside the race to develop safe, effective COVID-19 vaccines and novel COVID-19 therapeutics. This action raises resource allocation questions. If research funders have a certain amount of resources to spend on COVID-19 research, how much should be spent on testing existing medicines versus developing novel medicines versus developing vaccines versus conducting other essential non-biomedical research? In pandemics and epidemics, while biomedical research is important, non-biomedical research is also critical to:

- assess the well-being impact of the disease and public health policies, especially on those considered disadvantaged or marginalized prior to and due to the epidemic/pandemic; and
- assess whether social and economic policies mitigate identified negative well-being impacts and/or leave anyone made vulnerable behind in terms of well-being.

Fair allocation of resources would call for resources to be allocated to all four types of research, though it is possible that the lifesaving effects of biomedical research outputs, given that no vaccines or medicines existed for COVID-19, would justify more resources being spend on such studies. How much should be spent on vaccines versus novel medicines versus existing medicines needs further consideration, with criteria such as "likelihood of intervention efficacy," "cost-effectiveness of research," and "magnitude of benefit" being important factors to inform resource allocation (Pratt et al., 2018).

In the context of specific trials testing existing medicines like HCQ, remdesivir, and Kaletra for efficacy in treating COVID-19, it is also critical to ensure that participants represent the diversity within a country. Ideally, such trials will also be part of global efforts that capture the diversity of the world's population. However, fair subject selection means that those considered marginalized and disadvantaged should not be overly recruited relative to other populations. In the Australian context, this would mean that, for example, students, members of the Aboriginal and Torres Strait Islander population, and individuals from lower-income groups would not be disproportionality represented. Anthropological evidence shows that clinical trials in high-income countries primarily recruit those who are less well-off (Fisher, 2009). It is highly likely this could be exacerbated in the pandemic contexts, particularly as many workers have lost their jobs. It is also important to ensure that studies testing existing medicines are not disproportionately happening in low-income countries relative to richer countries.

Summary of Chapter

This chapter demonstrates why nations have established laws that govern the misuse of genetic information to the detriment of an individual when utilized for purposes outside the scope of the initial research (Genetic Non-Discrimination Act S.C.

2017, c.3) and the European Union (EU) (EU Charter of Fundamental Rights, Article 21: Non-discrimination). Genetic material collected from an individual that is found to contain coding patterns that may be later identified as associated with the increased risk for disease cannot be used as a reason for discrimination for employment or insurance purposes. While providing some protection to the individual, when communities or tribes are identified through genetic testing as possessing a genetic trait that might be considered undesirable, cultural and class discrimination is much more difficult to regulate yet may have equally as devastating impacts on the individual. Inclusion of a cultural representative in the study design phase as well as IRB inclusion of cultural expertise can help guide researchers to avoid study conduct and outcomes that may be abusive. Participant understanding of both consent and autonomy are critically important to ensure appropriate protection of study participants and their health information.

One Health (CDC, 2022) emphasized the interdependent nature of human, animal, and environmental health. Globalization has increased both the actual and potential spread of infectious diseases rapidly across borders. LSD is an example of a disease that is endemic in regions of Africa resulting in significant agricultural and economic limitations. When LSD threatened the Middle East and European agricultural markets, international cooperation in research and data sharing facilitated the prevention of LSD from becoming endemic in these regions. While reasonable efforts to distribute the benefits of research across partners, such as data sharing, are actively practiced, prevention of an invasive disease in middle- and high-income countries does not inform remediation of the same disease in low-income countries where it is endemic. Justice would dictate that African countries where LSD is endemic and that provide tissue samples for the development of strategies to treat and prevent an economically devastating disease would receive substantial benefit in disease treatment and prevention in exchange.

Finally, COVID-19 further expands the discussion about principles of justice and standards of evidence. Concerns for justice extend from local patients losing access to needed daily therapies for chronic conditions such as lupus, to potential global shortages of unproven medical therapies leading to the inability to effectively treat HIV with proven anti-retrovirals. The use of unproven therapeutics not only puts the individual taking the medication at a risk for adverse effects but can also put populations at risk due to drug shortages. In addition to justice arguments, under emergency circumstances, limited reported of positive test results limits the ability of scientists, physicians, and regulators to effectively evaluate the utility of medication across population due to a lack of contrasting studies that may demonstrate limitations not included in the positive finding studies.

Studies involving biospecimens and genetic tissue introduce an additional layer of complexity that needs to be considered in study design and continued throughout the study consent through completion of a study, monitoring for unintended consequences. Scientific advancements on previously collected specimens may inform our understanding without the equivalent consent that would be required if the study were undertaken anew today. Therefore, particular attention and sensitivity must be applied to studies that may result in additional insight in years and decades ahead.

Chapter Exercises

1. Draft a consent form that details information about biospecimen collection. Share and discuss your form with peers.
2. Reflect on the impact of unintended consequences from future discoveries of stored genetic information. Who should have access to the information? Why? What public decisions should be considered with these data?
3. Compare the U.S. Common Rule and a document of its equivalent from another country or region. Describe the guidance provided in the documents with respect to storage and use of biospecimens.
4. Locate a scholarly, peer-reviewed article on tissue and genetic code ownership. Summarize the issues discussed in the article. How would you share this information with research participants? How does the information link to your field of study or discipline?

9 Use of Technology

As technology progresses, so too does the potential for ethical challenges in research and other spaces. Continued enhancements in computing capacity allow large databases of information to be stored and mined, providing opportunities to analyze more trends across populations than could be envisioned when foundational guidance, such as the Helsinki Declaration (World Medical Association, various dates 1964–2008) and the Belmont Report (U.S. Department of Health, Education & Welfare, 1979) were developed. Today, researchers may even use data from social media (e.g., Twitter, Facebook) to track health outcomes such as outbreaks of influenza (Alessa & Faezipour, 2018).

Cloud storage has streamlined research data collection in many instances but raises concerns from potential participants about security and privacy protections. Similarly, the ease of data storage and transfer to other researchers can mean that additional research is undertaken without an explicit consent process with those whose data are stored (Rothstein, 2015).

Many large data sources used for health research, particularly social media databases, were not developed for research and, therefore, are not subject to the same kinds of oversight that occur with research reviewed by an ethics board (Rothstein, 2015). Thus, individuals whose data are included have typically not provided consent for research. Of particular concern is the situation where multiple data resources are linked allowing either (1) re-identification of an individual based on the combination of information from disparate sources (Rothstein, 2015), or (2) making sensitive data public as a result of linking one data source with another more public-facing data source (CIOMS, 2016).

The Council for International Organizations of Medical Sciences (CIOMS), jointly established by the WHO and UNESCO, provides ethical guidance for developing public health research and policy. Specifically, their guidelines caution that when researchers use online information, they should "anticipate, control, monitor and review interactions with their data across all stages of the research" (CIOMS, 2016). These guidelines also advise that just because data are not legally restricted from use in research, it does not relieve researchers of their ethical duties to protect the privacy and autonomy of participants (CIOMS, 2016). Similarly, the Regulatory and Ethics Work Group for the Global Alliance for Genomics

DOI: 10.4324/9781003033233-9

and Health provided a framework, developed by varied stakeholders, that can guide responsible and ethical conduct of research that involves data sharing. The framework is "guided by the human rights of privacy, non-discrimination and procedural fairness," and provides specific principles to facilitate responsible data sharing: transparency; accountability; engagement; data quality and security; privacy, data protection and confidentiality; risk–benefit analysis; recognition and attribution; sustainability; education and training; and accessibility and dissemination (Knoppers, 2014).

The twenty-first century has made the world smaller and more connected, creating opportunities for expanded research collaborations. Differences in the level of ethical review across various geographic boundaries may require investigators to take on more responsibility for responsible conduct of research, particularly when working with researchers in countries that may not yet have well-established research regulations or review committees (Krogstad et al., 2010; Makhoul et al., 2018).

The case studies in this chapter consider the following topic areas:

- privacy and data security considerations related to online provision of a mental health intervention;
- establishing trust regarding digital data collection and the potential for sharing electronic databases with other entities; and
- technological literacy and access to technological tools used in research.

Case Studies and Analyses

The case studies in this chapter are based in three countries: China, Nigeria, and the U.S. They deal with privacy considerations related to the provision of a mental health intervention through an online modality during the COVID-19 pandemic (China); finding culturally appropriate ways of creating and maintaining trust between researchers and community members when using digital data collection tools (Nigeria); and using technology for education during the informed consent process for participation in newborn genetic screening (U.S.). Through these case studies, we see how (a) online intervention delivery creates possibilities to reduce participant privacy and may affect the data collection process, (b) different levels of community engagement may impact the success of participant recruitment into studies that use unfamiliar technology, and (c) technology can be used for the informed consent process but may also create access barriers.

The case study analyses are provided by researchers in three countries: Kenya, Argentina, and Nigeria. The case study analysis from Kenya provides general principles relating to privacy and confidentiality relative to the use of the internet for telehealth delivery. This analysis also points out how interactions between researchers, information technology personnel, and participants (i.e., teamwork) may improve understanding of both technical and ethical components of a study. The analyses from Argentina and Nigeria point out the need for community engagement that addresses both the use of the technology and the infrastructure required to be able to consistently use technology for research. Given a range of familiarity with and access to tools, as well as infrastructure to support the use of technology for research, different community engagement approaches may be required for successful incorporation of technology for informed consent and data collection. In fact, some countries may simply not be able to apply technological approaches until more infrastructure to support the use of technology is available. In all of the case studies, we see the importance of including clear discussions of privacy, confidentiality, and data usage that are informed by extensive community engagement during study development.

Case Study #1 (China): Organizing Different but Related Online Groups to Manage Ethical Risks from Social Work Online Interventions during the COVID-19 Epidemic by Simin Peng and Huiqin Zhong

Context

In 2020, due to the risk of infection during the novel coronavirus epidemic, social distancing and shelter-in-place policies were implemented to prevent the spread of the virus. Social isolation broke the normal order of life. People stayed at home for lengthy periods of time and suffered emotional distress. Conflicts among family members at home occurred more frequently than before. Patients with a confirmed mental health diagnosis and people who have close contact with these patients often felt anxious and fearful, and needed guidance on how to deal with this circumstance.

In response, social workers developed new protocols for delivering services by internet that needed to be evaluated relative to established protocols. One such protocol involved online delivery of psychological counseling, provision of social support, and information about health. Participants included confirmed COVID-19 patients with mild symptoms, quarantined persons and their families, and community residents. The goals of online service delivery were to improve psychological adjustment to quarantine, to increase adoption of appropriate psychological coping tools, and to enrich family life.

Statement of the Issue

Internet service delivery can be convenient, overcoming the problem of providing services to those who live distant from a therapist, and also allowing a therapist to quickly organize experts from all over the country to share their knowledge and skills. However, there are also challenges to address, such as privacy and cybercrime. Privacy concerns may include the privacy of information shared in this setting with other group members, as well as the potential for those in the household to hear information that would otherwise have been shared in private. Cybercrime concerns may include potential hackers who could join and disrupt the online group or introduce spyware or virusware into group communications.

Outcomes

Online platforms for conducting virtual meetings have rapidly improved the security of their platforms. However, a therapist or a researcher who leads such groups still needs to address privacy within the context of a group therapy session, as well as the additional challenge of individuals participating from home where other household members may hear the discussion or introduce distractions by wanting the attention of the person who is participating in the therapy session. Such distractions may mean that the therapist or researcher does not receive the full information from the participant but could also mean that the participant does not fully understand the guidance being given by the therapist. Either situation may result in less effective outcomes than could be achieved in an in-person session. The development of practice guidelines for online network-based therapy or research sessions would be helpful for therapists or researchers and could establish a way to formally assess the benefits and harms of this format for information delivery. From a research perspective, it is important to determine whether network-based group therapy is comparable to in-person group therapy.

Questions to Consider

1. How can researchers and therapists ensure that information collected during online network-based sessions is kept private?
2. In the informed consent process, how can you distinguish between a person's usual online risk and risk from joining a study about online network-based counseling?
3. What organizations should be consulted during protocol development to ensure that participants' information from online groups does not make them potential targets for cybercrime?
4. How would you make sure that individuals have appropriate access to the opportunity to join an online network-based group?
5. How would you measure and ensure the effectiveness of network-based social work interventions relative to other ways of delivering the intervention?

Case Study Analysis: A Perspective from Kenya by Edna N. Bosire

Context

In this case study, the following issues were noted for an intervention delivered to a group using virtual meeting technology:

- Patients or participants sharing spaces with their families. This issue presents a scenario where provision of care or meetings is jeopardized given that the patient may have other people around them at home. In this case, maintaining privacy and confidentiality for group members can be challenging. Also, data collection may be influenced by guidance from someone in the home who provides context for questions or supplies potential responses to the participant, thus altering the data collected; and

- Concerns around cybercrime, which entails the possibility of disrupting the online meetings or introducing spyware or virusware into group communications.

Another issue to consider when using an online tool for health service delivery is thinking through how language and culture may impact the conversation, particularly when the intervention is delivered in a group model. Finally, informed consent may be more difficult to assess when verbal cues and body language may be less obvious when using a virtual platform.

This context provides an opportunity to discuss some of the ethical considerations in Kenya related to group-based health intervention delivery and data collection. For this discussion, the word telehealth is used to encompass both the delivery of the intervention (i.e., therapy provided in a group setting using virtual meeting technology) and the collection of data from participants.

Country-Specific Regulations

Informed Consent

Participants must be provided with enough information regarding their participation to promote informed decisions. This information should be provided in easy-to-understand formats or local languages that are commonly used, in order to promote understanding. In Kenya, the official languages are English and Swahili, but depending on where the participants are from, it might be more appropriate to conduct the study in one of the many local languages. The fact that there are many may require researchers to consider holding groups in different languages. An important element of informed consent is that participant rights should be respected in determining whether or how their personal data may be collected, used, or shared. Researchers should ensure they inform participants about what data will be collected, who will be collecting the data, how the data will be stored or managed, and when and how the data will be shared.

Privacy and Confidentiality

In Kenya, privacy and confidentiality is highly regarded and recommended by therapists, practitioners, and researchers and is stipulated in the national guidelines for telehealth (Ministry of Health—Kenya, 2017). Overall, regulations for telehealth in Kenya require that measures are put in place to ensure patient's data or information is protected against unintended access (Ministry of Health—Kenya, 2017). Researchers and information technology staff managing these data need to be informed how to protect the confidentiality of health data and must sign non-disclosure agreements.

Ensuring privacy could include (1) communicating with the patient/participant before the virtual meeting to determine whether they are in the right context and physical space for telehealth and providing suggestions that can promote privacy for a telehealth interaction (e.g., using a private room separate from the rest of the house or selecting a time when others are not home); and (2) using pseudonyms during the conversation. Further efforts related to reporting the findings include anonymizing individual data (i.e., using an unrelated identifier), and aggregating data to report group summaries to minimize the possibility of tracing data back to an individual participant.

Ensuring confidentiality requires the development of strategies to store de-identified participant information for research, and procedures to maintain the security of these data (e.g., limiting access to data through the use of username and password authentication, implementing restrictions so that members of the research team access only the data required for their job, and providing training to researcher team members on ways to ensure the security of data). In particular, researchers who also have a clinical role should be provided clear guidance on whether research data are part of the clinical record.

In addition to these general guidelines, ensuring confidentiality in the context of a telehealth intervention should be addressed with all group members. The researcher should inform group members about their role in maintaining the confidentiality of what is discussed by not sharing the conversation with others outside of the telehealth intervention. If the telehealth intervention is longitudinal, violations of confidentiality should be addressed with individual members, and may require that some people switch to individualized telehealth and be dropped from the study if they are unable to respect these restrictions to ensure confidentiality of others in the group.

Security

To avoid cyberattacks or crime, tools for virtual meetings must ensure that participants' data are handled in a secure manner by putting in place mechanisms that will guarantee privacy, confidentiality, data integrity, and provide documentation of both transmission and receipt of information (i.e., non-repudiation) at all times. Thus, the systems must be secure from unauthorized access, use, disclosure, disruption, modification, perusal, inspection, recording, and destruction. The data must be secure both in transit and when archived. When linking clinical data from other sources to the research records, researchers must come up with security tools that facilitate transmission of participants' data over the telecommunication networks, i.e., for web-based data access, or emails containing sensitive or private information. For example, use of Transport Layer Security (TLS) ensures that communications between a web browser and a server are private, the identity of the communicating parties can be authenticated, and data integrity is assured by use of digital certificates.

Cultural Considerations

It is important to ensure that telehealth is provided in ways that are culturally acceptable to participants. For example, in Kenya, there are over 47 ethnic communities. All these communities have diverse cultures and beliefs, which inform how participants view health and illness. It is important for researchers to ensure that they are conversant with participants' cultures and languages in order to align the communications and data collection along participants' belief systems. It is also highly recommended that researchers should come from participants' cultures or communities, or at least should demonstrate cultural and structural competence skills (Metzl & Hansen, 2014). This understanding is one way of promoting a positive relationship between researchers and participants, and will require attention to reaching out to communities to gain their trust, identify potential co-researchers, and establish appropriate remuneration to local researchers if they join the research team.

Researcher and Participant Competence in Using Technology

In Kenya, training staff on providing health care through online platforms has been a great success (Njenga et al., 2021). This training has been made possible by the fact that information technology is highly advanced in Kenya. People use various online platforms such as WhatsApp and Zoom to communicate. Similarly, researchers and participants must have the skills and expertise to use an online platform, and organizations should provide training on best practices to ensure their researchers' competence. This expectation also includes ensuring that both researchers and participants have access to equipment such as computers or smart phones, and have sufficient bandwidth to have successful virtual interactions.

Teamwork

Working in teams is one way to ensure success in telehealth types of research. The required skills are complex and may include input from those with information technology expertise, subject-matter expertise in the intervention and/or the outcomes being evaluated, research ethics and research methodology, and knowledge of local culture(s). Communication among group members can lead to new understanding about ethical considerations and technological practices. In cases where researchers are unable to adhere to ethical principles or experience technological challenges, team members should discuss this within the team to identify solutions. Technological challenges, in particular, may need to be referred to the relevant local supervisor and/or the national coordination team (Ministry of Health—Kenya, 2017).

Measuring the Effectiveness of a Network-Based Social Work Intervention

This case study used telehealth as a way to deliver a social work intervention to a group of research participants. As with any new venture, evaluation of process and outcomes is essential to defining what works, improving quality, and comparing participant outcomes relative to existing intervention delivery method(s). Some elements to consider are:

- Accessibility of the online platforms—participants struggling with getting the right gadgets, data or skills may not join or may drop out over time, and this may signify failure of the research process.
- Acceptability of the online interaction—participants should find both the provision of information (i.e., informed consent) and the quality of the intervention acceptable. The number who accept to be part of the telehealth intervention may signify success of the virtual sessions, but surveys or focus groups will provide additional information about acceptability of the interaction and barriers to participation.
- Demand for telehealth interventions—increased requests for telehealth reflect positively on elements of accessibility and acceptability. Eliciting feedback through satisfaction surveys or interviews may also provide information about ways to improve quality.
- Outcomes—determining whether there are there statistically important differences in participant outcomes will help to establish the equivalence of telehealth versus other methods of delivery for social work interventions.

Addressing Ethical Issues Involving Research Participants

Overall, provision of health interventions online is a great way to respond to challenges of accessing in-person interventions, especially during a crisis such as the COVID-19 pandemic when physical distancing was implemented to protect health and safety. However, ethical and technological aspects must be considered to ensure that access and intervention delivery maintain standards for responsible conduct of research. Researchers need training on applying ethical practices to the telehealth modality to ensure adequate structure for obtaining informed consent, maintaining privacy of participation, and maintaining confidentiality and integrity of data. Successful telehealth implementation offers the potential to expand and improve access to research participation for members of diverse communities, thus addressing health inequities that exist due to access barriers. In addition to the privacy and confidentiality considerations noted, key challenges may be having access to devices capable of participating in telehealth and having adequate bandwidth to ensure both participation and secure information transfer.

Case Study #2 (Nigeria): Ethical Issues in Research when Using Digital Data Collection Tools by Patrick L. Umeh

Context

A group of public health professionals conducted a study in Ohafia Local Government of Abia State, Nigeria. The purpose of the study was to understand the perceived effects of dry gin consumption among family men in selected rural communities of Ohafia. It is a long tradition of men in the 26 communities in Ohafia to welcome their guests with a bottle of locally brewed Schnapp (dry gin). As a result, most households keep a couple of bottles at home in anticipation of guests. Whenever men in the communities are gathered for any form of discussion including family, community, group discussions etc., their custom demands that dry gin be kept in their midst. Due to this tradition, people in these communities are generally perceived to consume a lot of dry gin. Dry gin is used as a token of respect for elderly people and distribution of the product is increasingly common. The study team, from the urban areas, collaborated with the opinion leaders of the selected communities to administer questionnaires to a random sample of men who live in a family setting within the selected communities.

Digital data collection tools installed on Android devices were used to administer the questionnaires. This tool use was to ensure quality data collection and to avoid unnecessary data cleaning. While the researchers chose digital data collection to ease the process, the majority of the participants seeing a digital device in the first community visited were reluctant to participate. Even though the purpose of the study and use of the data had been explained to them, some of the participants believed their information on a digital device meant their story would be reported on television. Many others said they didn't want to appear on the computer. The study team, after seeing the barriers associated with using this digital data collection tool, decided to use paper questionnaires to collect data from the next community. The team observed that participants in the second community did not express any concerns about their personal information being published on television or social media.

Statement of the Issue

The use of digital data collection tools in one rural community influenced whether individuals felt comfortable to participate in the study. Whereas the intended purpose of the digital data collection was to minimize the processes of data cleaning and to speed up data collection, some of the participants believed it exposed them to public shame. This perception presented the researchers as people with ulterior motives and threatened the trust between participants and researchers.

Outcomes

Based on the observation that respondents were more comfortable to participate with paper-based data collection, the study team decided to include a discussion in the informed consent of how digital data would be used. Those who had the purpose of digital data collection explained to them before being recruited to the study were far more relaxed about their participation than those in the first community who learned about digital data collection at the time of data collection. Further, some participants were guided to input the data directly into the device themselves. It was observed that interacting with participants about the devices and even letting them use the device increased their confidence in it. Eventually, digital data collection became totally acceptable in these communities. The team decided to go back to the communities that objected to digital data collection and offered explanation as part of the informed consent process before administering the questionnaire.

The study team noted a significant increase in the acceptance of the digital data collection tool among the initial communities, after adjusting the informed consent process and data collection protocol to inform participants about digital data collection.

Questions to Consider

1. How would you identify the protections and risks of digital data collection to a potential study participant?
2. Identify elements of community engagement that were used by the study team in this situation. What other approaches might they have considered to encourage community participation?
3. How might the study team interact with the participants after the completion of the study to confirm that data were used as specified in the consent document?

Case Study Analysis: A Perspective from Argentina by Felicitas Holzer and Alahí Bianchini

Context

A study aiming to collect data with digital tools among men in rural areas was conducted in Ohafia, Nigeria. Engagement with community leaders prior to study implementation and a conventional informed consent process to convey information about the use of digital tools were not successful in the first community approached. Potential participants did not adequately grasp the functioning of the Android devices to administer the questionnaires and data protection policies, even though they were clearly explained to them. As reported in the study, some of the participants believed that their information on a digital device implied their story would be reported on television. The ethical issues at stake comprise, on the one hand, the adequate involvement of study participants, and specifically their full and voluntary informed consent with a special focus on data privacy; and on the other hand, issues related to adequate community engagement in the sense that not only opinion leaders of communities, but also study participants must engage with researchers and fully understand the purpose of digital tools. The general approach to ensure ethical conduct is to ensure better communication with individual study participants up front.

Country-Specific Regulations

According to the Argentinian Civil and Commercial Code (Código Civil y Comercial (CCyC) article 59), each researcher is required to adequately inform research subjects or legal representatives about all important aspects of, generally speaking, medical acts, which includes clinical research, and the conditions of participation. Informed consent furthermore requires that all necessary information is presented *in written form* in the informed consent document and that a signature can only be obtained if the signing party has the competence to understand the information. Research subjects and/or their legal representatives are free to refuse participation. Informed consent is necessary to guarantee the ethical conduct of research, which is owed to the intimate relationship with a participant's dignity, integrity, and autonomy (ANMAT, 2006; Lamm, 2017). Personal information must be protected.

Argentina is an upper-middle-income country with a considerably heterogeneous population regarding ethnicity, education levels, and income. However, the degree of digitalization differs considerably between regions. There is a considerable degree of cultural and economic disintegration of indigenous people, and generally less familiarity with research procedures involving human subjects. Furthermore, people in rural areas often lack access to digital devices and new technologies. Hence, researchers potentially face similar problems in rural communities in Argentina and the use of digital tools to collect data ought to be included in the informed consent form. Specifically, there should be discussion of ways to protect participant privacy and confidentiality of the data collected, as well as who will access the data and what will be done with each data element. Wherever possible, data summaries should be used rather than reporting of individual data, and this level of detail should also be included in the informed consent form.

Addressing Ethical Issues Involving Research Participants

There are a number of issues that need to be considered in the research context in Argentina.

Risks of Digital Data Collection to Potential Study Participant

Generally speaking, proactive steps should be taken to avert possible harmful effects related to data collection, which may include peer review of research proposals, screening of participants, and the willingness to discontinue or modify research in progress, if necessary, to protect participants from unforeseen harm. Research participants can feel compromised in their dignity, self-esteem or sense of agency, or even feel distress regarding the use of sensitive data. In the particular case of the research study, it is evident that the consumption of alcohol among family men is a sensitive topic and that adequately dealing with data is very important to protect participants from stigmatization in (other) communities, as well as from the loss of self-esteem and autonomy to decide to whom their story should be revealed.

Elements of (Better) Community Engagement

Being confronted with the participants' reluctance toward digital data collection tools, researchers in Nigeria rethought and improved their community engagement strategy. The study team included a discussion about how digital data would be used in the informed consent form, and participants who received more extensive explanation of the purpose of digital data collection before being recruited to the study were less reluctant compared to participants confronted with digital data collection methods only at the time data collection was beginning. Letting participants input their data themselves also helped to increase acceptance among the participants.

Additionally, to attain greater acceptance of digital data collection tools, the communication strategy could have involved communities as a whole right from the beginning, and not only opinion leaders. Communication strategies may comprise community consultation and engagement events in which digital data collection tools would be explained, followed by a Q&A session and/or the opportunity to use the Android devices and become more familiar with them.

Post-Study Community Engagement

It is important that participants have the means to contact researchers once the study is finished. According to best practices, participants should receive the informed consent page with the contact information of the researchers before signing for the study. Researchers may re-contact participants and explain the research study findings to them once the data are analyzed or upon publication of the study. Again, this is an opportunity to reassure participants that their sensitive data and personal stories are not misused. As part of community engagement best practices, researchers are supposed to adequately address post-trial benefits owed to community members, such as sharing study findings. This may be done through sharing publications, in-person presentation of findings with Q&A, and/or generation of a brief report that is written at the appropriate level of health literacy, and ideally prioritizes infographics over text to convey information.

Enhanced community engagement was key to helping community members become familiar with digital data collection tools. Once this was done, community members were comfortable participating in digital data collection. In order to maintain trust, researchers who want to have an ongoing relationship that allows additional research, should consider that sharing study findings after the completion of the research is part of their responsibility to the community. Ideally, this sharing should be specified in the informed consent form.

Case Study #3 (United States): Ensuring Equal Access to a Technological Intervention by Erin P. Johnson and Erin Rothwell

Context

Current regulations regarding the need to collect consent from new parents for the use of newborn screening (NBS) residual dried bloodspots in research is in flux. National attention in the U.S. to this issue is illustrated by recent congressional debate regarding the renewal of the "Newborn Screening Saves Lives Act," which supports the maintenance and growth of state-run newborn screening programs nationwide (U.S. Congress, 2019). The rationale for delaying the renewal of this long-standing and important legislation is that congressional leaders could not come to consensus regarding the need to obtain consent for research with residual

NBS dried bloodspots (Office for Human Research Protections, 2018a, 2018b). Many states have stopped retaining NBS bloodspots for research use because of the ethical and technical challenges of obtaining consent for population-based research. Low consent rates in birthing centers or hospitals may be the result of an inefficient and ineffective consent process rather than decisions made by parents to decline permission. U.S. researchers designed a study to determine the best way to educate women about newborn screening, the potential use of the residual dried bloodspot for use in research, and what it means to consent to this research use. Previous studies completed by the researchers indicated that women struggle with comprehending pamphlet-provided education in the days after delivery of the baby. Reasons for lower comprehension included the hectic environment of the hospital after delivery, as well as the abstract nature of unspecified research using stored biospecimens. Although providing a pamphlet with a consent form can meet the regulatory requirements for informed consent, it may not be the most effective way of providing the information needed to decide on participating in NBS research. As such, technology-based options, such as use of videos and interactive apps, delivered in the hospital soon after the birth of a child may be an appropriate solution to these consent challenges by providing a more engaging informed consent process.

Statement of the Issue

The difficulty in providing consent education via technology is access to the technology itself. How does one ensure that all parents are able to access education through a cell phone, tablet, or computer? Many hospitals do not have funding to support the technology needed for these interventions, requiring parents to use their own devices, when available. Further, there are a number of cultural and historical concerns, health literacy issues, and ensuring adequate comprehension during the postnatal environment that raise additional challenges to the informed consent process and that technology platforms may not address.

Outcomes

The U.S. researchers determined that access to the educational intervention through technology would need to be variable to ensure demographic inclusion, with participants opting in to the method that best suited their needs rather than being assigned to a particular technological approach. Participants would only be randomized to the research condition (tablet-based video or app) after they had indicated their preferred technological approach (paper- or tablet-based) (Rothwell et al., 2020). Randomizing after participants chose the manner in which they accessed the information allowed the researchers to ensure that the access method was distributed across trial groups. While this increases potential error due to self-selection into how and when the educational tool is accessed, this real-life approach allows for all participants to receive the necessary education rather than limiting the study.

Questions to Consider

1. How can parental permission be obtained for the use of leftover dried bloodspots from newborn screening that promotes a more informed decision by parents?
2. What are the ethical differences between an opt-out and an opt-in approach for research use of leftover dried bloodspots by parents?
3. How would technological literacy impact the use of apps or videos in low resource settings?

Case Study Analysis: A Perspective from Nigeria by Eunice Bajomo

Context

A technology-based approach for the use of newborn screening (NBS) in Nigeria would raise a set of ethical challenges for the country and consideration of the use of NBS for research could not be addressed immediately. Throughout the entirety of the country, there are a series of interconnected problems such as widespread inequality in health-care access, crumbling infrastructure and severely under-supported medical staff, cultural differences, and regional destabilization that all impact the ability to effectively incorporate this technology. Probably the most overarching issue in Nigeria is a lack of adequate infrastructure in many of the government-funded hospitals. With the lack of significant government funding supporting hospitals, many basic amenities are unavailable; for example, many hospitals do not have access to 24-hour electricity. With this inability to maintain an essential utility, the attempt to include technology-based consent from mothers into the already strained health-care system in Nigeria would encounter a significant barrier. Furthermore, the lack of sufficient funding leads to frequent strikes from unpaid health-care workers, continuing the strain on an already overburdened system. As a result, the majority of health care is accessed by only the wealthiest Nigerian citizens in private hospitals. Therefore, access to medical care is almost completely unavailable to those outside of this class, stressing a lack of equity and equality for all citizens outside this group. Literacy levels also remain below 50 percent in the population, which can make it difficult to implement just one method of obtaining consent from mothers. These various access issues result in an uneven distribution of potential to participate in the informed consent process, and thus in research, and contribute to health disparities across the Nigerian population.

Religion also presents further ethical concerns. The two major religions in Nigeria are Islam and Christianity, with the former being practiced primarily in the north of the country and the latter in the south. The country as a whole is a very religious nation. There are multiple sects of Christianity that have varying views towards possible NBS testing. For example, some Christian objections to NBS concern the view that children are created in God's image, and that any outside actions or procedures that arise from technological sources, such as the identification of genetic anomalies found during NBS, interfere in the natural life cycle and interventions taken would violate tenets of their religion. In the north, along with the existing issues presented in Christian communities in the country, several other influences impact the implementation of NBS in general. African Islamic extremist groups, most notably Boko Haram and the West African Province branch of the Islamic State, have fought extensively against Western education and technology in the region. This view has led to extreme destabilization and unrest in some parts of the north, severely limiting access to both medical care for women and to a technology-based consent process as an entry point to research. These restrictions further limit the population of Nigerian mothers who would be able to participate in research similar to that proposed in the case study.

Country-Specific Regulations

The central limitation inherent in the implementation of a technology-based approach to NBS in Nigeria lies with the severely underfunded and understaffed health-care infrastructure; this challenge may be experienced in other low- and middle-income countries (LMICs). Trying to incorporate new strategies in an already overburdened system is practically impossible. This issue is further exacerbated by religious objections, cultural differences, destabilized regions in the north of the country, and high rates of illiteracy. With a lack of understanding by the public on what NBS is, how testing occurs, and its potential to benefit communities, it is essential to first establish a foundational understanding of NBS that responds to these various barriers of access to proper medical care.

Addressing Ethical Issues Involving Research Participants

Three main issues come to mind relative to the research proposed in the case study. First, because medical care is predominantly available only to the wealthiest Nigerian citizens, addressing the ethical principle of justice is important in any research that uses medical facilities as a point of recruitment into studies. Second, the high technology approach to informed consent is currently not an option in Nigeria, and health systems strengthening efforts would be needed to address the technology requirements, as well as the literacy challenges. Finally, cultural considerations must be addressed in order for the health systems strengthening activities to be accepted by people with different understandings about health, health-care delivery, and research participation. These issues are addressed in the following paragraphs.

Justice

The principle of justice requires that benefits and burdens of research be equally distributed across those who participate in research. Thus, the selection of research participants requires an evaluation of this equal distribution of benefits and burdens. In application, "injustice arises from social, racial, sexual and cultural biases institutionalized in society" (Department of Health, Education, and

Welfare, 1979). In Nigeria, such an injustice can arise when a study recruits participants only from those who seek maternity care at a medical facility because this generally includes only those with high socio-economic status. Thus, only a portion of the citizenry participate in the study, making it non-representative of the country as a whole. Further, any benefits of participating in research would accrue only to this group, representing the social injustice of this method for participant selection. In this case study, only the wealthy would have access to the information provided by NBS and the ability to make timely decisions about ways to mitigate the negative impacts of genetic disorders, contributing to a health disparity by socio-economic status for genetic conditions such as inborn errors of metabolism.

Health Systems Strengthening

The World Health Organization provided a framework for health systems strengthening that has six building blocks (World Health Organization, 2007). Among these building blocks is a commitment to equitable access of needed services, such as maternity care. Other building blocks have to do with workforce adequacy, flow of information, and adequate funding, as addressed in this section. This need for health systems strengthening as a stepping stone to including research in the clinical setting is not unique to Nigeria, and similar efforts may be required in other LMICs before a technology-based informed consent process would even be feasible.

As noted in the contextual information for this case analysis, the workforce in medical facilities is inadequate to provide the care that is needed in Nigeria. Changing this situation will require country-level funding and policy to train and compensate the necessary workforce. In addition to this investment, additional training will be needed on topics such as NBS, particularly for current employees who may not have received this information during their own training. This strategy of building a foundation of understanding about NBS among care providers is focused on starting from the beginning; that is, establishing a network of care that generates greater understanding of NBS as the foundation for obtaining patient consent to participate in such a screening program and its associated research.

Beyond training medical staff on NBS, a large effort will be required to disseminate this information directly to the public and build trust between community members and medical professionals. Outreach campaigns and community engagement programs will be necessary to create the trust to engage in formal health care and will set the stage for introducing opportunities to participate in research which can inform best practices for health care. With the goal of increasing access to medical resources and procedures for all Nigerians, there should also be measurable improvements in health equity metrics and related health outcomes. From a research perspective, increasing access to medical care can also improve justice when trusted medical facilities become an access point for participating in research that has the potential to improve health outcomes.

Cultural Considerations

Training staff members about NBS is an opportunity to begin identifying cultural and religious concerns that may be associated with NBS. Similarly, implementing community outreach will provide opportunities to further identify and understand cultural and religious concerns that reflect the perspectives of local community members. The concerns about NBS that arise from different religious and cultural perspectives will require including community voices in the process of developing educational materials. Because many of these perspectives are rooted in religious beliefs, emphasizing the benefits of NBS in the context of religious doctrine is an important consideration. Similarly, decision-making is rooted in cultural contexts. For example, in the Islamic faith, men are at the center of the family's decision-making. Thus, outreach and education should not just target women, but should also have components that engage with men to explain the benefits of NBS, and the cultural congruence between a screening approach to better health and religious practice or doctrine.

Due to the high rates of illiteracy across the country, educational materials should include not just written information, but also visual illustrations and auditory approaches to sharing information. Talking points for common questions will help to standardize the educational content delivered, which should enhance trust between medical professionals and community members. Educational materials that engage people from different cultural backgrounds and with different learning styles emphasize inclusivity and further support efforts to achieve justice.

The country of Nigeria (and many other LMICs) is not currently ready to implement routine NBS, nor a technology-based approach to obtaining informed consent as one option for engaging patients in health research. The use of medical facilities as a recruitment source for research studies is one that is frequently used all over the world, but Nigeria's health infrastructure challenges limit access to care—and potential research opportunities—to those of high socio-economic status. This limitation raises concerns about the principle of justice in making research opportunities, and their potential benefits, available to all citizens. Moving Nigeria closer to greater inclusivity in research through improving access to medical care will involve a country-level commitment to health systems strengthening that can have benefits for both individual and community health.

Summary of Chapter

In summary, research that incorporates technology for informed consent, delivery of an intervention, and/or collection of data should address, through informed consent, issues that are related to: (i) participant privacy, (ii) data storage and means of protecting confidentiality, (iii) how data will be analyzed, and (iv) how stored data may be shared and used in the future. The case studies and analyses in this chapter all point out that the feasibility of using technology in research depends on literacy, infrastructure support, access to the technology, and familiarity of participants and researchers with the capabilities of the technology. Successful implementation, thus, is reliant on adequate stakeholder engagement in both the planning and implementation of the research.

The case study from China used an online platform to deliver a telehealth intervention. This study raises issues related to privacy and confidentiality during group interactions, where the informed consent process should provide guidance to participants about ways to ensure these principles are upheld. Nevertheless, the possibility that there are other people off-screen who may hear private information during the session or interact with the participants and potentially alter their responses cannot be ruled out. These risks should be explicitly addressed in the informed consent process, with suggestions of ways participants can respect privacy of the group members and keep confidential the information that is shared in the group setting. The analysis from Kenya provides guidance on ways to address participant privacy and confidentiality, as well as the data security and storage issues that are raised when delivering telehealth interventions.

In the second case study, the issue of working with participants who are not familiar with technology was raised. Even though the technical infrastructure was available, and researchers were well versed in the cloud-based data collection platform, participant unfamiliarity with technology was a barrier to study participation in Nigeria. Despite community engagement with leaders prior to starting the study, participants were skeptical that their privacy would be respected and suggested that their data would be shared in ways that would identify and shame them. A similar situation can be found in Argentina, where technical capacity for online data collection is available, but there are some communities where access to technology is limited and participants would be concerned about online data collection and storage. Both sets of researchers acknowledged the need for a deeper level of engagement in these situations, that could include explanations of the protections put in place by the researchers and opportunities for potential participants to interact with the technology in low stakes ways and gain their buy-in prior to the research data collection. The case study analysis further points out that the informed consent process—which should include explanations about participant privacy, data confidentiality, and data use now and into the future—is in effect from the initial recruitment through reporting back to participants what the study findings were. Such a long-term commitment builds trust between community members and researchers and is important for meaningful, ongoing relationships that advance the health of individuals and communities.

The third case study contrasts a high-tech environment (U.S.) with a low-tech environment (Nigeria). In this case study, participants were randomized to paper-based compared to technology-delivered informed consent processes, with the goal of standardizing and improving the informed consent process in the busy post-natal environment of the hospital (Rothwell et al., 2020). The case study analysis provides insight on the challenges of implementing a highly technological approach to research in medical facilities in many LMICs. These challenges include intermittent availability of electricity, lack of equipment to deliver the technology and/or inadequate bandwidth, inadequate number of health-care workers to conduct research, and the need for training on research topics and protocols. These challenges will likely mean that highly technological approaches—while holding promise for improving access to research and addressing issues of justice where only select groups currently receive the benefits of research—must wait until infrastructure challenges can be addressed.

This chapter emphasizes the importance of defining the processes for protecting participant privacy and how data will be collected, stored, and used in a secure and ethical manner. Researchers implementing technological means for intervention delivery, data collection, and storage will benefit from broad stakeholder input on the technical approaches, as well as the cultural nuances that must be addressed in order to gain buy-in regarding the use of technology in research.

Chapter Exercises

1. The elements of privacy and confidentiality in research that uses technology may require additional explanation; additionally, technology may change over the course of the study. How would you ensure that privacy and confidentiality are adequately represented during the informed consent process? Include consideration of format (written, visual, and auditory) to address potential health literacy issues.
2. Develop a list of issues you could consider when addressing cultural differences between researchers and participants in a study that relies on technology during interactions with participants.
3. In your own research field, who would you consider as essential stakeholders to include prior to study implementation? Design a list of questions you could use to gather their input about the use of technology for data collection and storage, and the cultural concerns you might need to address during informed consent.
4. How does access to technology facilitate the principle of justice in research? How does it act as a barrier?

10 Brief Conclusions

As readers consider the information in this book, we think it is quite clear that there is no universal method for how to identify and manage ethical dilemmas in the research space. There are geographically and culturally nuanced ways that ethical issues are defined and addressed, which makes contemplation of a "one-size-fits-all" approach a bit difficult. Researchers from different countries have aspirations of ensuring that research has integrity and is conducted responsibly; however, there are many methods of determining what the issue is and social, political, and economic components that guide how researchers and ethics panels make decisions about what is allowed in the research context and how to protect human subjects in these endeavors. The issues highlighted in this book can be categorized into three areas: shifting to a view of "what should be done" as a researcher or research team, involving community stakeholders in research to provide clarity for cultural expectations, and creating documents that clearly highlight what a study is about and how human subjects are to access these study activities and understand what it means to them.

As we reviewed the case studies and the analyses of these studies, one idea that become clearer to us was to develop a set of ethical guidelines that might enable a global shift from what should not be done to *what should be done* such that the concept of responsible conduct of research becomes a more proactive and positive approach to supporting research integrity. This shift would be away from the Nuremberg Code language of what researchers "should not do" to considerations of "what is right" to do given the cultural and social contexts in which research takes place. Given this stance in today's political and social climate, the involvement of community stakeholders in all aspects of the research endeavor could (1) assist in broadening the understanding of research ethics by these community stakeholders and (2) promote hearing different voices along the research continuum.

In support of this, the U.S. Belmont Report and the U.N. Declaration of Helsinki both describe expectations about the treatment of human subjects such that they are treated fairly, seen as autonomous beings, and are not harmed in research activities. Although best practices on a global scale may be difficult to concisely define, the ideas about what researchers should do would have a similar base for all researchers that could then be applied with cultural, political, economic, and geographic nuance.

DOI: 10.4324/9781003033233-10

As the case studies illustrate, researchers and ethics review panels need to know their "audiences"—community stakeholders and research participants. Thus, ensuring that voices from the wider community, not just scientists or research teams, are part of the research landscape could promote a clearer understanding of what a study is about and ensure that participants are supported as they accept being involved in research activities.

Documents used for studies are crucial not only because they provide written evidence of what a study is about, but they also provide information to potential participants, funders, ethics review panels, and the community at large. As noted throughout the case studies and case analyses, information can be highlighted in these documents; however, they are only useful for those who can read the language and/or are reading at the language level of the documents. Thus, they provide ideas, but for those who are not literate or are not literate about research there can be misunderstanding between researchers and participants, which may increase the potential harm or reduce the fairness of a study. Exploration of alternatives to written consent may need to be considered for some potential participants. Along these lines, the issues identified in these chapters should be used as a template for discussion of options to build a positive culture with research teams and with study participants.

There are many ethical areas in research that this book did not address, i.e., specific populations (e.g., people with disabilities), the constantly changing landscape of artificial intelligence, use of "big data," genomics and de-identification of an individual's information, the management and secure storage of data, and data sharing. Several topics were mentioned briefly, but additional exploration of these topics is warranted. These areas are part of the continuously changing domains of research ethics and responsible conduct of research. More research is needed in these areas and a second book about dilemmas would be useful. In summary, we hope the information in this book will be a stepping stone for students and their faculty to delve into common areas for discussion in developing studies that promote research integrity.

Appendix: Table of Videos by Chapter, Topic, and Country/Region of Focus

Chapter	Video Link	Topic	Country/Region of Focus
2: Minors	https://www.nih.gov/health-information/nih-clinical-research-trials-you/personal-stories	Children's accounts of being involved in research and why it is important	United States
	http://www.childrenandclinicalstudies.org/Why-Its-Important/The-Documentary	Personal accounts of families with children in clinical studies	United States
	https://healthtalk.org/clinical-trials-parents-experiences/involving-children-in-decisions-child-assent	Several short videos of child assent	United Kingdom
	https://www.unicef-irc.org/research/ethical-research-and-children/	Ethical research involving children	UNICEF
	https://ukhealthcare.uky.edu/patients-visitors/clinical-trials	Clinical trials at United Kingdom Health Care	United Kingdom
	https://cambridgebrc.nihr.ac.uk/public/taking-part-research/	Why children are asked to take part in research studies	United Kingdom (Cambridge)
	https://methods.sagepub.com/video/an-introduction-to-the-ethics-of-research-with-children	Introduction to ethics of research with children	United Kingdom (Lancaster)
	https://www.youtube.com/watch?v=Pmsj1-0K7Kc	Participatory research with children	Germany
	https://www.childresearch.net/projects/crn_asia/2022_09.html	Child research network Asia	Asia
3: Females	https://www.hopkinsmedicine.org/henriettalacks/index.html	Henrietta Lacks' story	United States
	https://www.youtube.com/watch?v=2S24Is1ezNA&feature=emb_imp_woyt	Drug trials using only male participants	United States

	https://www.cnn.com/2014/02/04/world/americas/canada-brain-dead-woman/index.html	Doctors keep brain-dead woman alive for her fetus	Canada and United States
	https://globalhealthtrainingcentre.tghn.org/articles/webinar-community-engagement-clinical-research-involving-pregnant-women/	Research involving pregnant women in Malawi	Malawi
	https://www.youtube.com/watch?v=FcecqQkyKoM	Lack of diversity in research (TED Talk)	Sub-Saharan Africa
	https://www.ted.com/talks/alyson_mcgregor_why_medicine_often_has_dangerous_side_effects_for_women/transcript	Why medicine often has dangerous side effects for women (TED Talk)	Various
	https://recruit.ucsf.edu/diversity-research-participation-why-its-important	Importance of diversity in research participants	United States
	https://www.youtube.com/playlist?list=PLhhb-JA5bQ7PSbv78gkphGNMZjzu-Y8r-	Innovating for maternal and child health in Africa (IMCHA)	Africa
	https://www.youtube.com/watch?v=z33HUk8ZBqs&list=TLGGtyaltzLPpysxNzA2MjAyMg	Gender & COVID-19 research	Asia/Pacific
	https://www.youtube.com/watch?v=oWjbxH8iV64	Maternal health	Latin America
4: Males	https://www.dmu.ac.uk/research/research-faculties-and-institutes/health-and-life-sciences/centre-for-lgbtq-research/centre-for-lgbtq-research.aspx	Introducing a center that focuses on inclusive research	United Kingdom
	http://www.worldbank.org/en/topic/socialdevelopment/publication/investing-in-a-research-revolution-for-lgbti-inclusion	Advocates for research on social inclusion of LGBTI population	World Bank Group

(Continued)

Chapter	Video Link	Topic	Country/Region of Focus
	https://www.abc.net.au/news/2016-04-26/prep-trials-to-expand-in-queensland-after-brave-government-call/7360710	Drug trial for PrEP to reduce fears and normalize use	Australia
	https://www.youtube.com/watch?v=HgpAV3B0IDQ	Mass incarnation of black men and lack of African Americans in clinical studies	United States
	https://www.youtube.com/watch?v=TUmmig7BZJw	Doctors discuss men's health	Southeast Asia
5: Language	https://www.nala.ie/health-literacy/	Health literacy	Ireland
	https://www.youtube.com/watch?v=phcrWpHOxMY&feature=emb_imp_woyt	Health literacy and health communication	United States
	https://www.youtube.com/watch?v=LWzOfWCoRAU	Invention of medical translator Medibabble	United States
	https://www.youtube.com/watch?v=-x6DLqtaK2g	Health Literacy (TED Talk)	United States
	https://www.healthliteracyeurope.net/hls-eu	Health literacy	Europe
	https://www.uea.ac.uk/groups-and-centres/global-research-translation-award-project/family-literacy	Family literacy	Ethiopia, Philippines, Malawi
	https://www.youtube.com/watch?v=saeU_dK14ok	How is tradition hindering health literacy	Kenya, Tanzania, Malaysia
6: Participant Incentives	https://www.ted.com/talks/boghuma_kabisen_titanji_ethical_riddles_in_hiv_research	HIV research in Sub-Saharan Africa	Sub-Saharan Africa
	http://www.accounting.uci.edu/support/guides/research-participants.html	How a university compensates research participants	United States

	https://blogs.lse.ac.uk/africaatlse/2014/05/14/incentivising-health-workers-in-zambia/	Incentivizing health workers	Zambia
	https://www.povertyactionlab.org/evaluation/designing-incentives-combat-urban-diabetes-india	Incentives in research projects	India
	https://www.youtube.com/watch?v=fMzIEFPKy8E	Research with vulnerable & underrepresented populations	United States
	https://videocast.nih.gov/Summary.asp?File=16194&bhcp=1	Fair subject selection & recruitment	United States
7: Mental Health	https://www.pcori.org/events/2016/pcori-and-nami-briefing-patient-centered-mental-health-research	Briefing about current work being done to better mental health research	United States
	https://www.nimh.nih.gov/news/media/2018/mobile-technology-for-global-mental-health-research.shtml	Using technology/telehealth to improve mental health research globally	Globally (but hosted by the NIH in United States)
	https://www.nimh.nih.gov/archive/media/2017/treatment-targets-target-engagement-and-target-populations-in-mental-health-services-research-to-improve-public-health-examples-from-the-field	Child mental health (case studies from Uganda, Kenya, Ghana, South Africa)	Global
	https://www.nimh.nih.gov/news/media/2020/identifying-new-directions-in-mental-health-disparities-research-innovations-with-a-multidimensional-lens-day-one-part-one	Disparities on mental health research	United States

(*Continued*)

Chapter	*Video Link*	*Topic*	*Country/Region of Focus*
	https://r4hc-mena.org/our-work/	Mental health programs to address research gap in mental health research	Palestine and Lebanon
	https://www.youtube.com/watch?v=Nybusx0uk8s	Youth Mental Health Challenge	UNICEF
	https://www.psych.ox.ac.uk/news/ethics-in-mental-health-digital-innovations-for-young-people-in-africa-digital-campaign	Ethics in mental health digital innovations for young people in Africa	United Kingdom and Africa
	https://www.youtube.com/watch?v=uKd9MxBzAUc	Chains of mental health in West Africa	West Africa
8: Genetics and Biospecimens	https://www.ted.com/talks/paul_knoepfler_the_ethical_dilemma_of_designer_babies/transcript#t-42311	The ethical dilemmas of designer babies	Various
	https://www.tedmed.com/talks/show?id=309672	Personal, social, and policy consequences of genomic sequencing	United States
	https://www.ted.com/talks/jennifer_doudna_we_can_now_edit_our_dna_but_let_s_do_it_wisely/transcript?language=en#t-5855	Ethics of gene editing	United States
	https://www.ted.com/talks/chris_nowinski_can_i_have_your_brain_a_quest_for_truth_on_concussions_cte/transcript	Brain Bank for the study of Chronic Traumatic Encephalopathy (CTE) and using donation cards to pledge brain to science after death	United States
	https://www.ted.com/talks/keolu_fox_why_genetic_research_must_be_more_diverse/transcript	Importance of including indigenous, non-European people in genetic research	United States

	https://www.ted.com/talks/tania_simoncelli_should_you_be_able_to_patent_a_human_gene/transcript	Ethics of patenting human genes	United States
	https://www.smh.com.au/healthcare/biggest-biobank-in-the-southern-hemisphere-to-revolutionise-medical-research-in-nsw-20171113-gzk5os.html	Biobank	Australia
	https://www.youtube.com/watch?v=bV8T-D1GWFU&feature=emb_imp_woyt	Map the Gap	South Asia
	https://www.youtube.com/watch?v=YU8ZoNp3rlc	High-depth African genomes	South Africa
	https://www.youtube.com/watch?v=TLyky0Db6hU	Medical genomics	Latin America to the Pacific
	https://www.youtube.com/watch?v=A6gFtS8VMEU	Gene editing	Latin America
	https://www.youtube.com/watch?v=QcPyDNS0Joc	Epigenetics & Genomics	Middle East and South Asia
9: Use of Technology	https://www.ted.com/talks/nicholas_christakis_how_social_networks_predict_epidemics	Using social networks to predict epidemics	United States
	https://www.ted.com/talks/john_wilbanks_let_s_pool_our_medical_data/transcript#t-53033	Argument in favor of data sharing (genetic and medical records)	United States
	https://www.ted.com/talks/malte_spitz_your_phone_company_is_watching/transcript?referrer=playlist-the_dark_side_of_data	Data Retention Directive of the European Union and privacy issues with big data	Germany
	https://www.ted.com/talks/paul_root_wolpe_it_s_time_to_question_bio_engineering/transcript?referrer=playlist-new_tech_new_morals#t-97510	Lack of ethical guidelines governing the proper use of bio-engineering	

References

23andMe (2019). Research consent document-key points. Retrieved November 29, 2022, from www.23andme.com/about/consent/

Afgan, E., Baker, D., van den Beek, M., Blankenberg, D., Bouvier, D., Čech, M., Chilton, J., Clements, D., Coraor, N., Eberhard, C., Grüning, B., Guerler, A., Hillman-Jackson, J., Von Kuster, G., Rasche, E., Soranzo, N., Turaga, N., Taylor, J., Nekrutenko, A., & Goecks, J. (2016). The Galaxy platform for accessible, reproducible and collaborative biomedical analyses: 2016 update. *Nucleic Acids Research* 44(W1), W3–W10. https://doi.org/10.1093/nar/gkw343

Aküzüm, T., Çakmak, B., & Gökalp, Z. (2010). Evaluation of water resources management in Turkey. *International Journal of Agricultural and Natural Sciences* 3(1), 67–74.

Alaoui, M.I.. Association Al Hayat-Chaines de Vie. National newborn resuscitation training program for midwives. Collaboration with the American LDS-Charities, 2014–2017 Rabat, Kingdom of Morocco.

Alessa, A., & Faezipour, M. (2018). A review of influenza detection and prediction through social networking sites. *Theoretical Biology & Medical Modeling* 15, 2. doi:10.1186/s12976-017-0074-5

American College of Obstetricians and Gynecologists (2015). Ethical considerations for including women as research participants. Retrieved November 27, 2022, from https://www.acog.org/clinical/clinical-guidance/committee-opinion/articles/2015/11/ethical-considerations-for-including-women-as-research-participants?utm_source=redirect&utm_medium=web&utm_campaign=otn

American Psychiatric Association (APA) Task Force on Research Ethics. (2006). Ethical principles and practices for research involving human participants with mental illness. *Psychiatric Services* 57(4), 552–557.

Amnesty International (1999). Pakistan: Violence against women in the name of Honor. Retrieved August 13, 2022, from https://www.amnesty.org/en/documents/ASA33/017/1999/en/

Anam, F. (2019). Representations of women's role in Pakistan: A critical analysis through drama serials. Retrieved August 13, 2022, from https://vc.bridgew.edu/jiws/vol20/iss3/2/

Ancestry (2019). Privacy philosophy. Retrieved November 29, 2022, from https://www.ancestry.com/c/privacyphilosophy.

ANMAT (2006). Boletín para profesionales. Retrieved November 3, 2020, from http://www.anmat.gov.ar/webanmat/Publicaciones/Boletines/Profesionales/Boleprof_Agosto_2006.pdf

Aronowitz, R. (2014). "Screening" for prostate cancer in New York's Skid Row: History and implications. *American Journal of Public Health* 104(1): 70–76.

Atıcı, E. (2009). About regulation on clinical research. *Uludağ Üniversitesi Tıp Fakültesi Dergisi* 35(3), 147–152.

Baldor, L.C. (2021). Reversing Trump, Pentagon releases new transgender policies. Military Times. Retrieved August 5, 2022, from https://www.militarytimes.com/news/pentagon-congress/2021/03/31/reversing-trump-pentagon-to-release-new-policy-for-transgender-people-to-serve-in-the-military/

Barmania, S., & Aljunid, S.M. (2016). Navigating HIV prevention policy and Islam in Malaysia: Contention, compatibility or reconciliation? Findings from in-depth interviews among key stakeholders. *BMC Public Health* 16, 524. https://doi.org/10.1186/s12889-016-3247-y

Barstow, C. (2020). 150% Principle. Retrieved July 4, 2022, from https://rightuseofpower.org/150-principle/

BBC (2014). Ethics: A general introduction. Retrieved April 24, 2022, from http://www.bbc.co.uk/ethics/introduction/intro_1.shtml

Beskow, L. (2016). Lessons from HeLa Cells: The ethics and policy of biospecimens. *Annual Reviews of Genomics & Human Genetics* 17, 395–417.

Bharath, H. (2019). Indian initiatives aim to break science's language barrier. *Nature*. Retrieved May 22, 2022, from https://www.nature.com/articles/d41586-019-01815-1

Bhutta, Z. (2002). Ethics in international health research: A perspective from the developing world. *Bulletin of the World Health Organization* 80, 114-120. Retrieved November 27, 2022, from https://www.scielosp.org/scielo.php?pid=S0042-96862002000200006&script=sci_arttext&tlng=en

Biden, J.R. (2021). Executive Order on enabling all qualified Americans to serve their country in uniform. President Actions. The White House. USA. Retrieved March 9, 2022, from https://www.whitehouse.gov/briefing-room/presidential-actions/2021/01/25

BKK-VBU-Meine Krankenkasse (n.d.). Sozialdienst des Krankenhauses. Retrieved May 22, 2022, from https://www.meine-pflegekasse.de/wegweiser/anlaufstellen/sozialdienst-des-krankenhauses/

Black, R.E. (2001). Micronutrients in pregnancy. *The British Journal of Nutrition* 85(Suppl 2), S193–S197. https://doi.org/10.1079/bjn2000314

Brall, C., Berlin, C., Zwahlen, M., Egger, M., & Vayena E. (forthcoming manuscript, 2020). Attitudes, beliefs, concerns, and expectations regarding personalised health research and biobanking among the Swiss population – a national large-scale survey.

Budge, S.L., Tebbe, E.N., & Howard, K.A.S. (2010). The work experiences of transgender individuals: Negotiating the transition and career decision-making processes. *Journal of Counseling Psychology* 57, 377–393. https://doi:10.1037/a0020472

Büken, N. (2017). Children as participants of clinical research. *Journal of Pediatric Infectious Disease* 11(2), 76–83. https://doi.org/10.5578/ced.201720

Calistri, P., Clercq, K. D., Gubbins, S., Klement, E., Stegeman, A., Abrahantes, C., Marojevic, D., Antoniou, S.-E., & Broglia, A. (2020). Lumpy skin disease epidemiological report IV: Data collection and analysis. *EFSA Journal* 18(20), e06010. doi:10.2903/j.efsa.2020.6010.

Calistri, P., DeClercq, K., de Vleeschauwer, A., Gubbins, S., Klement, E., Stegeman, A., Cortiñas Abrahantes, J., Antoniou, S., Broglia, A., & Gogin, A. (2018). Lumpy skin disease: Scientific and technical assistance on control and surveillance activities. *EFSA Journal* 16(10). https://doi.org/10.2903/j.efsa.2018.5452

Calvo, P. (2009). Vacuna experimental: multan por un millón de pesos a responsables de un ensayo con niños. Clarín. Retrieved July 4, 2022, from https://www.clarin.com/ultimo-momento/vacuna-experimental-multan-millon-pesos-responsables-ensayo-ninos_0_B1sgDHKA6Kl.html

Cans, C., De-la-Cruz, J., & Mermet, M.A. (2008). Epidemiology of cerebral palsy. *Paediatrics and Child Health* 18(9), 393–398. doi:10.1016/j.paed.2008.05.015.

Catt, S., Fallowfield, L., Jenkins, V., Langridge, C., & Cox, A. (2005). The informational roles and psychological health of members of 10 oncology multidisciplinary teams in the UK. *British Journal of Cancer* 93, 1092–1097. https://doi.org/10/1038/sj.bjc.6602816

CensusInfo India (n.d.). Final Population Totals 2011. Retrieved November 27, 2022, from https://byjus.com/free-ias-prep/census-of-india-2011-key-facts-and-data/

Centers for Disease Control (CDC) (2022) One Health Basics. Retrieved July 12, 2022, from https://www.cdc.gov/onehealth/basics/index.html

Chaara, I. (2018). Moroccan mother's religiosity: Impact on daughters' education. In D.H. Gray & N. Sonneveld (Eds.), *Women and social change in North Africa: What counts as revolutionary?* (pp. 143–165). Cambridge: Cambridge University Press.

Chakraborty, S., Chopra, M., Mani, K., Giri, A., Banerjee, P., Sahni, N., Siddhu, A., Tandon, N., & Bharadwaj, D. (2018). Prevalence of vitamin B12 deficiency in healthy Indian school-going adolescents from rural and urban localities and its relationship with various anthropometric indices: a cross-sectional study. *Journal of Human Nutrition and Dietetics* 31(4), 513–522. https://doi.org/10.1111/jhn.12541

Chakrabarty, S., Kabekkodu, S.P., Brand, A., & Satyamoorthy, K. (2016). Perspectives on translational genomics and public health in India. *Public Health Genomics* 19, 61–68. doi:10.1159/000442518

Chao, M. & Chu, Z. (2009). Beijing Chaoyang District people's court, civil judgements, No. 02608. Blog. SINA. Retrieved May 22, 2022, from http://blog.sina.com.cn/s/blog_c2b037dd01011f29q.html.

Chatfield, K., Lightbody, C.E., Qayum, I., Ohly, H., Rasgado, M.C., Watkins, C., & Lowe, N.M. (2022). Can an ethics code help to achieve equity in international research collaborations? Implementing the global code of conduct for research in resource-poor settings in India and Pakistan. *Research Ethics* 18(4), 281–303.

Cheff, R. (2018). Compensating research participants: A survey of current practice in Toronto. Retrieved July 4, 2022, from https://www.wellesleyinstitute.com/publications/compensating-research-participants-a-survey-of-current-practices-in-toronto/

Chetoui, A., Kaoutar, K., El Kardoudi, A., & Boutahar, K. (2018). Epidemiology of diabetes in Morocco: Review of data, analysis, and perspectives. *International Journal of Scientific and Engineering Research* 9(12), 1310–1316.

Chilcot, J., Wellsted, D., & Farrington, K. (2011). Illness perceptions predict survival in haemodialysis patients. *American Journal of Nephrology*, 33(4), 358–363. https://doi.org/10.1159/000326752

Chope, R.C., & Strom, L.C. (2008). Critical considerations in career and employment counseling with transgender clients. In G.R. Walz, J.C. Bleuer, & R.K. Yep (Eds.), *Compelling counseling interventions: CelebratingVISTAS' fifth anniversary* (pp. 125–135). Alexandria, VA: American Counseling Association.

Choudhary, S. & Sharma, R. (2021). 70 years on, India is still fighting over a national language. *The Times of India*. Retrieved May 22, 2022, from https://timesofindia.indiatimes.com/india/70-years-on-india-is-still-fighting-over-a-national-language/articleshow/86196740.cms.

CMS (2007). Routine costs in clinical trials. Centers for Medicare & Medicaid Services. Retrieved May 22, 2022, from https://www.cms.gov/medicare-coverage-database/details/ncd-details.aspx?NCDId=1&fromdb=true

CMS (2022). Clinical Laboratory Improvements Amendments (CLIA). Centers for Medicare & Medicaid Services. Retrieved August 1, 2022, from https://www.cms.gov/Regulations-and-Guidance/Legislation/CLIA

Coaston, J. (2019). The intersectionality wars. Vox, May 28, 2019. Retrieved July 7, 2022, from https://www.vox.com/the-highlight/2019/5/20/18542843/intersectionality-conservatism-law-race-gender-discrimination

Cochrane Training. (2019). Patient and public involvement in research: What, why and how? Retrieved July 2, 2022, from https://training.cochrane.org/resource/patient-and-public-involvement-research-what-why-and-how

Codigo Civil y Comercial de la Republica Argentina. Bibliotecadigital.com.ar. Retrieved November 16, 2020, from http://www.bibliotecadigital.com.ar/CCyC/index.htm?context=220

Comision Coordinadora del Sector de Justicia (2009). Ley de Protección Integral de la Niñez y Adolescencia (LEPINA). Retrieved May 22, 2022, from https://escuela.fgr.gob.sv/wp-content/uploads/Leyes/Leyes-2/LEY-DE-PROTECCION-INTEGRAL-DE-LA-NI%C3%91EZ-Y-ADOLESCENCIA.pdf

Common Rule (2012). Retrieved May 22, 2022, from https://www.govinfo.gov/content/pkg/CFR-2012-title21-vol1/pdf/CFR-2012-title21-vol1-sec50-20.pdf

Cong, Y. (2018) Legal and ethical issues of justice: Global and local perspectives on compensation for serious adverse events in clinical trials. In D. Schroeder, J. Cook, F. Hirsch, S. Fenet, & V. Muthuswamy (Eds.), *Ethics Dumping*. Springer, Cham: Springer Briefs in Research and Innovation Governance. https://doi.org/10.1007/978-3-319-64731-9_15.

Consejo Superior de Salud Pública (2021). Comite Nacional de Etica de Investigación en Salud (CNIS). Retrieved May 22, 2022, from http://cssp.gob.sv/servicios/descarga-de-documentos/comite-nacional-de-etica-de-investigacion-en-salud/

Constitution of the Republic of South Africa | South African Government. [online] Retrieved February 25, 2022, from: http://www.gov.za/DOCUMENTS/CONSTITUTION/constitution-republic-south-africa-1996-1

Council for International Organizations of Medical Sciences (CIOMS) (2016). International ethical guidelines for health-related research involving humans. Retrieved July 4, 2022, from https://cioms.ch/wp-content/uploads/2017/01/WEB-CIOMS-EthicalGuidelines.pdf

Council of Europe (1999). Convention for the protection of Human Rights and Dignity of the Human Being with regard to the Application of Biology and Medicine: Convention on Human Rights and Biomedicine. Retrieved May 22, 2022, from https://www.coe.int/en/web/conventions/full-list/-/conventions/treaty/164.

Croatian LSD Surveillance Programme in 2018. (2018). Veterinary and Food Safety Directorate Ministry of Agriculture, Croatia SCoPAFF Meeting. Retrieved January 21, 2022, from https://ec.europa.eu/food/system/files/2018-12/reg-com_ahw_20181211_lumpy-skin-disease_hrv.pdf

Dag Hammarskjöld University College of International Relations and Diplomacy in Zagreb, Kovačić, M.K., & Bencetić, D.B. (2016). The meat sector in Croatia. Retrieved November 29, 2022, from https://www.flandersinvestmentandtrade.com/export/sites/trade/files/market_studies/2016-Croatia-Meat-Sector.pdf

Davidson, A.J., & O'Brien, M. (2009). Ethics and medical research in children. *Pediatric Anesthesia* 19(10), 994–1004. https://doi.org/10.1111/j.1460-9592.2009.03117.x

De Sousa, A. (2008). Psychiatric issues in renal failure and dialysis. *Indian Journal of Nephrology* 18(2), 47–50. https://doi.org/10.4103/0971-4065.42337

Defend2020.eu. (2022). Defend2020.Eu. Retrieved November 29, 2022, from https://defend2020.eu/about-us/

Douglass, K., et al. (2021). Language diversity and challenges to communication in Indian emergency departments. *International Journal of Emergency Medicine* 14(1), 57. doi:10.1186/s12245-021-00380-7.

Dresser, R. (1992). Wanted: Single, white male for medical research. *The Hastings Center Reports* 22(1), 24–29. https://doi:10.2307.3562720

Dumon, O. (2017). In research fields, women make progress but there's still a gap. Retrieved November 27, 2022, from https://www.huffpost.com/entry/in-research-fields-women-make-progress-but-theres_b_58e26afae4b0d804fbbb751a

Ejaz, L., Chang, A.H., Bari, K., Rizvi, B.M. (n.d.). Role and Procedural Guidelines, Research Ethics Committee, Center for Business and Economics Research/School of Economics and Social Sciences/Institute of Business Administration – Karachi. Retrieved November 30, 2022, from https://www.iba.edu.pk/sess/pdf/research-ethics-guidelines-and-forms.pdf

Essery, R., Geraghty, A.W., Kirby, S., & Yardley, L. (2017). Predictors of adherence to home-based physical therapies: A systematic review. *Disability Rehabilitation* 39(6), 519–534. doi:10.3109/09638288.2016.1153160.

Eti İçli, G., Anıl, N. K., & Kılıç, B. (2016). The affecting factors of consumers' organic food purchase choices. *Kırklareli Üniversitesi İktisadi ve İdari Bilimler Fakültesi Dergisi* 5(2), 93–108.

European Food Safety Authority (EFSA), Calistri, P., de Clercq, K., Gubbins, S., Klement, E., Stegeman, A., Abrahantes, J.C., Marojevic, D., Antoniou, S., Broglia, A., European Food Safety Authority (EFSA), Calistri, P., de Clercq, K., Gubbins, S., Klement, E., Stegeman, A., Abrahantes, J.C., Marojevic, D., Antoniou, S., & Broglia, A. (2020). Lumpy skin disease epidemiological report IV: Data collection and analysis. EFSA. Retrieved November 29, 2022, from https://efsa.onlinelibrary.wiley.com/doi/10.2903/j.efsa.2020.6010

Fischer, B.A. (2006). A summary of important documents in the field of research ethics. *Schizophrenia Bulletin* 32(1), 69–80.

Fisher, J. (2009). *Medical research for hire: The political economy of pharmaceutical clinical trials*. Rutgers, NJ: Rutgers University Press.

Fleissig, A., Jenkins, V., Catt, S., & Fallowfield, L. (2006). Multidisciplinary teams in cancer care: Are they effective in the UK? *The Lancet Oncology* 7(11), 935–943. https://doi.org/10.1016/S1470-2045(06)70940-8

FoE Africa. (2007, March). Does Africa need a genetically modified solution to diarrhea? FoE Africa ALERT01. Retrieved November 27, 2022, from http://www.eraction.org/publications/medicinerice.pdf.

Foreman, K.J., Marquez, N., Dolgert, A., Fukutaki, K., Fullman, N., McGaughey, M., Pletcher, M.A., Smith, A.E., Tang, K., Yuan, C.-W., Brown, J.C., Friedman, J., He, J., Heuton, K.R., Holmberg, M., Patel, D.J., Reidy, P., Carter, A., Cercy, K., … Murray, C.J.L. (2018). Forecasting life expectancy, years of life lost, and all-cause and cause-specific mortality for 250 causes of death: Reference and alternative scenarios for 2016–40 for 195 countries and territories. *The Lancet* 392(10159), 2052–2090. https://doi.org/10.1016/S0140-6736(18)31694-5

Fraser, M.W., & Galinsky, M.J. (2010). Steps in intervention research: Designing and developing social programs. *Research on Social Work Practice* 20(5), 459–466. https://doi.org/10.1177/1049731509358424

Freese, B. (2007). *A grain of caution: A critical assessment of pharmaceutical rice.* Washington, D.C.: Center for Food Safety.

Ghana Statistical Service (GSS), Ghana Health Service (GHS), and Macro International. (2009). Ghana Maternal Health Survey 2007. Calverton, Maryland, USA: GSS, GHS, and Macro International.

Gibbs, P. (2021). Lumpy Skin Disease in cattle. Merck Veterinary Manual. Retrieved November 29, 2022, from https://www.merckvetmanual.com/integumentary-system/pox-diseases/lumpy-skin-diseasein-cattle?autoredirectid=16683

Goldstein, Z., Corneil, T.A., & Greene, D.N. (2020). When gender identity doesn't equal sex recorded at birth: The role of the laboratory in providing effective healthcare to the transgender community. *Clinical Chemistry* 63(8), 1342–1352. https://doi.org/10.1373/clinchem.2016.258780

Gonmei, Z., & Toteja, G.S. (2018). Micronutrient status of Indian population. *The Indian Journal of Medical Research*, 148(5), 511–521. https://doi.org/10.4103/ijmr.IJMR_1768_18

Grady, C. (2015). Institutional Review Boards. *Chest* 148(5), 1148–1155. doi:10.1378/chest.15-0706.

Gray, D.H. (2015). *Beyond feminism and Islamism – Gender and equality in North Africa.* London: I.B. Tauris.

Gren, L.H., Jaggi, R., Frost, C.J., & Benson, L.S. (2018). Reframing ethics when unintended consequences arise: A case report of a multiple micronutrient study during pregnancy in Ghana. *African Journal of Current Medical Research.* doi:http://dx.doi.org/10.31191/afrijcmr.v2i2.16

Grotlüschen, A., & Riekmann, W. (2011). Literalität von Erwachsenen auf den unteren Kompetenzniveaus. Retrieved May 22, 2022, from https://leo.blogs.uni-hamburg.de/wp-content/uploads/2011/12/leo-Presseheft_15_12_2011.pdf

Guttmacher Institute (2013). Abortion in Ghana. Retrieved November 27, 2022, from https://www.guttmacher.org/fact-sheet/incidence-abortion-and-provision-abortion-related-services-ghana

Hamal, M., Dieleman, M., De Brouwere, V., & Buning, T. (2020). Social determinants of maternal health: A scoping review of factors influencing maternal morality and maternal health service use in India. *Public Health Reviews* 41, 13–37. https://publichealthreviews.biomedcentral.com/track/pdf/10.1186/s40985-020-00125-6.pdf

Hannum, H. (1998). The UDHR in National and International Law. *Health and Human Rights* 3(2), 144–158.

Haseltine, W. (2020). Moderna's claim of favorable results in its vaccine trial is an example of 'publication by press release'. *Washington Post.* Retrieved November 29, 2022, from https://www.washingtonpost.com/opinions/2020/05/19/rush-share-good-news-covid-19-drugs-is-undermining-science/

Hatzenbuehler, M.L., & Pachankis, J.E. (2016). Stigma and minority stress as social determinants of health among lesbian, gay, bisexual, and transgender youth: Research evidence and clinical implications. *Pediatric Clinics of North America* 63(6), 985–997.

Hayırlıdağ, M., Arslan, M. F., & Örnek Büken, N. (2016). Ethicolegal arguments on genetically modified foods and evaluation of intercontinental situation. *Tarım Ekonomisi Dergisi* 22(1), 1–9.

Hina. (2019). Croatia declared free from Lumpy Skin Disease. Total Croatia News. Retrieved January 21, 2022, from https://www.total-croatia-news.com/business/33706-lumpy-skin-disease

History Channel (2010). Nuremberg Trials. Retrieved April 24, 2022, from https://www.history.com/topics/world-war-ii/nuremberg-trials

Homedes, N., & Ugalde, A. (Eds.) (2014). *Clinical trials in Latin America: Where ethics and business clash*. Cham: Springer.

Hoy-Ellis, C.P., & Fredriksen-Goldsen, K.I. (2017). The mental health of transgender older adults: General and minority stress. *American Journal of Community Psychology* 59(3–4), 295–305. https://doi.org/10.1002/ajcp.12138

Hoy-Ellis, C.P., Shiu, C., Sullivan, K.M., Kim, H.J., Sturges, A.M., & Fredriksen-Goldsen, K.I. (2017). Prior military service, identity stigma, and mental health among transgender older adults. *The Gerontologist* 57(suppl 1), S63–S71. https://doi.org/10.1093/geront/gnw173

Hungarian LGBT Alliance, Transvanilla Transgender Association, Hatter Society, Budapest Pride, & Labrisz Lesbian Association. (2015, September). LGBTQI Rights in Hungary. ILGA. Retrieved November 27, 2022, from https://ilga.org/downloads/Hungary_STAKEHOLDERS_REPORT.pdf

Hussain-Gambles, M., Leese, B., Atkin, K., Brown, J., Mason, S., & Tovey, P. (2004). Involving South Asian patients in clinical trials. *Health Technology Assessments* 8(42), 1–109. doi:10.3310/hta8420.

Huston, M., & Larsen, R.C. (2014). Occupational mental health & workplace violence. In J. Ladou & R. Harrison (Eds.), *CURRENT Occupational Medicine & Environmental Medicine* (5th ed.). New York: LANGE/McGraw.

ICCSR. ICSSR at a glance. Indian Embelam. (n.d.). Retrieved December 6, 2021, from https://icssr.org/icssr-glance

ICHGCP (n.d.). Essentielle Dokumente für die Durchführung einer Klinischen Prüfung. Retrieved May 22, 2022, from https://ichgcp.net/de/8-essential-documents-for-the-conduct-of-a-clinical-trial

India Census Bureau (2011). Abstract of speakers' strength of languages and mother tongues. Retrieved May 22, 2022, from https://censusindia.gov.in/2011Census/Language-2011/Statement-1.pdf].

India Government (n.d.). Know India. Retrieved May 22, 2022, from https://knowindia.india.gov.in/states-uts/

India State-Level Disease Burden Initiative Collaborators (2017). Nations within a nation: Variations in epidemiological transition across the states of India, 1990–2016 in the Global Burden of Disease Study. *The Lancet* 390(10111), 2437–2460. doi:https://doi.org/10.1016/S0140-6736(17)32804-0.

Indian Constitution (n.d.) Indian Constitution art. 343(1). Retrieved May 22, 2022, from https://indiankanoon.org/doc/379861/?__cf_chl_tk=704PofGU3RXI.6NBDk6EU_8WIH_ehsYEvZJkflE_BA4-1641169173-0-gaNycGzNB5E

Indian Council of Medical Research (1980). Policy statement on ethical considerations involved in research on human subjects. Retrieved November 27, 2022, https://ethics.ncdirindia.org/asset/pdf/Policy_Statement_1980.pdf

Indian Council of Medical Research (2006). Guidelines. Retrieved November 27, 2022, from https://main.icmr.nic.in/guidelines

Institut du Cerveau (n.d.). The Key Figures. Retrieved May 22, 2022, from https://icm-institute.org/fr/chiffres-cles

International Commission of Occupational Health (2019). About ICOH. Retrieved April 24, 2022, from http://www.icohweb.org/site/about-icoh.asp

Jafarey, A.M., Iqbal, S.P., & Hassan, M. (2012). Ethics review in Pakistan: The credibility gap. *Journal of the Pakistan Medical Association* 62(12), 1354–1357.

James, S.L., Abate, D., Abate, K.H., Abay, S.M., Abbafati, C., Abbasi, N., Abbastabar, H., Abd-Allah, F., Abdela, J., Abdelalim, A., Abdollahpour, I., Abdulkader, R.S., Abebe, Z., Abera, S.F., Abil, O.Z., Abraha, H.N., Abu-Raddad, L.J., Abu-Rmeileh, N.M.E., Accrombessi, M. M.K., … Murray, C.J.L. (2018). Global, regional, and national incidence, prevalence, and years lived with disability for 354 diseases and injuries for 195 countries and territories, 1990–2017: A systematic analysis for the Global Burden of Disease Study 2017. *The Lancet*, 392(10159), 1789–1858. https://doi.org/10.1016/S0140-6736(18)32279-7

Jat, T.R., Deo, P.R., Goicolea, I., Hurtig, A.K., & San Sebastian, M. (2015). Socio-cultural and service delivery dimensions of maternal mortality in rural central India: A qualitative exploration using a human rights lens. *Global Health Action*, 8, 24976. https://doi.org/10.3402/gha.v8.24976

Kadam, R., & Karandikar, S. (2012). Ethics committees in India: Facing the challenges! *Perspectives in Clinical Research* 3(2), 50–56. https://doi.org/10.4103/2229-3485.96444

Kamran, S.P. (2019). Women, art & politics in Pakistan: Rethinking feminism through feminist art. *PEOPLE: International Journal of Social Sciences* 5(2), 712–719. doi:10.20319/pijss.2019.52.712719

Kanter, J., Koh, C., Razali, K., Tai, R., Izenberg, J., Rajan, L., Van Griensven, F., & Kamarulzaman, A. (2011). Risk behaviour and HIV prevalence among men who have sex with men in a multiethnic society: A venue-based study in Kuala Lumpur, Malaysia. *International Journal of STD & AIDS* 22(1), 30–37. https://doi.org/10.1258/ijsa.2010.010277

Kaplan, R. (2001). The aversion project – Psychiatric abuses in the South African Defence Force during the apartheid era. *South African Medical Journal Forum* 91(3), 216–217.

Kidd, S.A., Howison, M., Pilling, M., Ross, L.E., & McKenzie, K. (2016). Severe mental illness in LGBT populations: A scoping review. *Psychiatric Services*, 67(7), 779–783. https://doi.org/10.1176/appi.ps.201500209

Kidd, S.A., Veltman, A., Gately, C., Chan, K.J., & Cohen, J.N. (2011). Lesbian, gay, and transgender persons with severe mental illness: Negotiating wellness in the context of multiple sources of stigma. *American Journal of Psychiatric Rehabilitation*, 14, 13–39. https://doi.org/10.1080/15487768.2011.546277

Kim, W.O. (2012). Institutional review board (IRB) and ethical issues in clinical research. *Korean Journal of Anesthesiology* 62(1), 3–12. doi:10.4097/kjae. 2012.62.1.3.

Kindergeld (2022). Kindergeld: Anspruch-Antrag-Höhe und Auszahlung. Retrieved May 22, 2022, from https://www.kindergeld.org/

Kıvılcım, Z. (2012). Cartagena Protocol and Turkey's biosafety legislation. *Marmara Avrupa Araştırmaları Dergisi*, 20(1), 99–121.

Klingstrom, T., Mendy, M., Meunier, D., Berger, A., Reichel, J., Christoffels, A., Bendou, H., Swanepoel, C., Smit, L., Mckellar-Basset, C., Bongcam-Rudloff, E., Soderberg, J., Merino-Martinez, R., Amatya, S., Kihara, A., Kemp, S., Reihs, R., & Muller, H. (2016). Supporting the development of biobanks in low- and medium-income countries. 2016 IST-Africa Week Conference, 1–10. https://doi.org/10.1109/ISTAFRICA.2016.7530672

Knoppers, B.M. (2014). Framework for responsible sharing of genomic and health-related data. *The HUGO Journal* 8, 3. doi:10.1186/s11568-014-0003-1

Konkel, L. (2015). Racial and ethnic disparities in research studies: The challenge of creating more diverse cohorts. *Environmental Health Perspective* 123(12), A297–A302. https://doi:10.1289/ehp.123-A297

Kotecha, P.V. (2008). Micronutrient malnutrition in India: Let us say "no" to it now. *Indian Journal of Community Medicine*, 33(1), 9–10. https://doi.org/10.4103/0970-0218.39235

Krogstad, D., Diop, S., Diallo, A., Mzayek, F., Keating, J., Koita, O.A., & Toure, Y.T. (2010). Informed Consent in International Research: The Rationale for Different Approaches. *American Journal of Tropical Medicine and Hygiene* 83(4), 743–747. doi:10.4269/ajtmh.2010.10-0014

Lamm, E. (2017). Consentimiento informado para actos medicos e investigaciones en salud. DELS. Retrieved November 3, 2020, from http://www.salud.gob.ar/dels/entradas/consentimiento-informado-para-actos-medicos-e-investigaciones-en-salud

Lancet, The. (2019). Raising the profile of men's health. *The Lancet* 394(10211), 1779. doi:10.1016/S0140-6736(19)32759-X

Legislation Information System (Mevzuat Bilgi Sistemi) (2013). İŞYERİ HEKİMİ VE DİĞER SAĞLIK PERSONELİNİN GÖREV, YETKİ, SORUMLULUK VE EĞİTİMLERİ HAKKINDA YÖNETMELİK. Retrieved May 31, 2020, from https://www.mevzuat.gov.tr/mevzuat?MevzuatNo=18615&MevzuatTur=7&MevzuatTertip=5(in Turkish)

Leighton, P. (2006). Study on infants in Peru sparks ethics inquiry. Retrieved May 22, 2022, from https://www.scidev.net/global/news/study-on-infants-in-peru-sparks-ethics-inquiry/.

Logan, M.W. & Link, N.W. (2019). Taking stock of drug courts: Do they work? *Victims and Offenders* 14(3), 238–298. doi:10.1080/15564886.2019.1595249

Lucksted, A. (2004). Raising issues: Lesbian, gay, bisexual, & transgender people receiving services in the public mental health system. National Alliance on Mental Illness New York State. Retrieved July 6, 2022, from https://www.naminys.org/images/uploads/pdfs/Multicultural%20Issues%20in%20Veterans%20Mental%20Health.pdf

Makhoul, J., Chehab, R., Shaito, Z., & Sibai, A. (2018). A scoping review of reporting 'Ethical Research Practices' in research conducted among refugees and war-affected populations in the Arab world. *BMC Medical Ethics* 19, 36. doi:10.1186/s12910-018-0277-2

Martin, E. (2001). *The woman in the body: A cultural analysis of reproduction*. New York: Beacon Press.

Mastroianni, A.C., Faden, R., & Federman, D. (1994). *Women and health research: Ethical and legal issues of including women in clinical studies: Volume 1*. National Academies Press. Retrieved November 27, 2022, from https://www.ncbi.nlm.nih.gov/books/NBK236535/

Mathur, R., & Swaminathan, S. (2018). National ethical guidelines for biomedical & health research involving human participants, 2017: A commentary. *The Indian Journal of Medical Research* 148(3), 279–283. https://doi.org/10.4103/0971-5916.245303

McCann, T.V., & Lubman, D.I. (2018). Stigma experience of families supporting an adult member with substance misuse. *International Journal of Mental Health Nursing* 27(2), 693–701. doi:0.1111/inm.12355

McGinley, M., & Horne, S.G. (2020). An ethics of inclusion: Recommendations for LGBTQI research, practice, and training. *Psychology in Russia: State of the Art*, 13(1), 54–69. https://doi.org/10.11621/pir.2020.0106

Mduluza, T., Midzi, N., Duruza, D., & Ndebele, P. (2013). Study participants incentives, compensation, and reimbursement in resource-constrained setting. *BMC Medical Ethics* 14(S4). https://doi.org/10.1186/1472-6939-14-S1-S4

Mentzell Ryder, P. (2015). Beyond critique: Global activism and the case of Malala Yousafzai. *Literacy in Composition Studies* 3(1), 175–187. doi:10.21623/1.3.1.14

Merriam-Webster Online Dictionary (2017). Global-Definition. Retrieved April 24, 2022, from https://www.merriam-webster.com/dictionary/global

Metzl, J.M., & Hansen, H. (2014). Structural competence: Theorizing a new medical engagement with stigma and inequality. *Social Science & Medicine* 103, 126–133. https://doi.org/10.1016/j.socscimed.2013.06.032

Meyer, I.H. (2003). Prejudice, social stress, and mental health in lesbian, gay, and bisexual populations: Conceptual issues and research evidence. *Psychological Bulletin* 129(5), 674–697.

Miklóstitle, E. (2006). *"Férfi a férfival, nő a nővel": Homoszexualitás a történelemben, a társadalomban és a kultúrában*. Budapest: Corvina.

Mills, E., Beyrer, C., Birungi, J., & Dybul, M.R. (2012). Engaging men in prevention and care for HIV/AIDS in Africa. *PLoS-Med* 9(2), e1001167.

Ministry of Health. (2017). Kenya Standards and Guidelines for mHealth Systems. Retrieved April 29, 2022, from https://www.health.go.ke/wp-content/uploads/2020/02/Revised-Guidelines-For-Mhealth-Systems-May-Version.pdf

Modi, N., Vohra, J., Preston, J., Elliott, C., Van't Hoff, W., Coad, J., … for a working party of the Royal College of Pediatrics and Child Health. (2014). Guidance on clinical research involving infants, children, and young people: An update for researchers and research ethics committees. *Archives of Disease in Childhood* 99(10), 887–891. https://doi.org/10.1136/archdischild-2014-306444

Murphy, S. (2018). The growth of medical cannabis in Europe. Health Europa. Retrieved November 27, 2022, from https://www.healtheuropa.eu/medical-cannabis/88296/

Nasrullah, M., Zakar, R., & Zakar, M.Z. (2014). Child marriage and its associations with controlling behaviors and spousal violence against adolescent and young women in Pakistan. *Journal of Adolescent Health* 55(6), 804–809. doi:10.1016/j.jadohealth.2014.06.013

National Center for Educational Statistics (2006). The health literacy of America's adults: Results from the 2003 National Assessment of Adult Literacy. Retrieved May 22, 2022, from https://nces/edu.gov/pubs2006/2006483.pdf.

National Commission for the Protection of Human Subjects of Biomedical and Behavioral Research. (1979). The Belmont report: Ethical principles and guidelines for the protection of human subjects of research. U.S. Department of Health and Human Services. Retrieved November 29, 2022, from https://www.hhs.gov/ohrp/regulations-and-policy/belmont-report/read-the-belmont-report/index.html

National Human Genome Research Institute (NHGRI) (2017). Human Subjects Research: Participation in Genomics Research. Retrieved November 29, 2022, from www.genome.gov/about-genomics/policy-issues/Human-Subjects-Research-In-Genomics

National Institutes of Health (NIH) (2017). Revisions to the Common Rule. Retrieved November 29, 2022, from https://www.genome.gov/about-genomics/policy-issues/Human-Subjects-Research-in-Genomics/Highlights-of-Revisions-to-the-Common-Rule

Nilsson, M.P., Emmertz, M., Kristoffersson, U., Borg, A., Larsson, C., Rehn, M., Winter, C., Saal, L.H., Brandberg, Y., & Loman, N. (2018). Germ-line Mutations in BRCA1 and BRCA 2 incidentally revealed in a biobank research study: Experiences from re-contacting mutation carriers and relatives. *Journal of Community Genetics* 9(3), 201–208.

Njenga, F., Ongeri, L., Nguithi, A., Anundo, J., Mugane, M., Kimari, Z., Mbugua, G., Kaigwa, C., & Atwoli, L. (2021). Integrating telepsychiatry services in a care setting in Kenya: A case report. *British Journal Psychiatry International* 19(2), E3. doi:10.1192/bji.2021.38

Nuffield Council on Bioethics (2003). The use of genetically modified crops in developing country: A follow-up Discussion Paper. Retrieved May 22, 2022, from https://www.nuffieldbioethics.org/publications/gm-crops-in-developing-countries.

Nuremberg Code (n.d.). Office of NIH History & Stetten Museum. Retrieved November 26, 2022, from https://history.nih.gov/display/history/Legislative+Chronology?preview=%2F1016866%2F30736589%2Fnuremberg.pdf

Nuremberg Military Tribunals (1949). Trials of War Criminals before the Nuremberg Military Tribunals under Control Council Law No. 10 (2), 181–182. Washington, D.C.: U.S. Government Printing Office. Retrieved July 4, 2022, from https://history.nih.gov/research/downloads/nuremberg.pdf

O'Mathuna, D.P. (2006). Human dignity in the Nazi era: Implications for contemporary bioethics. *BMC Med Ethics* 7, 1–12. doi:10.1186/1472-6939-7-2.

OECD/European Observatory on Health Systems and Policies. (2019). Hungary: Country Health Profile 2019, State of Health in the EU, OECD Publishing, Paris/European Observatory on Health Systems and Policies, Brussels. Retrieved July 6, 2022, from https://doi.org/10.1787/4b7ba48c-en

Office for Human Research Protections (OHRP) (2018a). Attachment C – Updated FAQs on informed consent for use of biospecimens and data. Retrieved June 29, 2022, from https://www.hhs.gov/ohrp/sachrp-committee/recommendations/attachment-c-faqs-recommendations-and-glossary-informed-consent-and-research-use-of-biospecimens-and-associated-data/index.html

Office for Human Research Protections (OHRP) (2018b). Newborn blood spot. Retrieved June 29, 2022, from https://www.hhs.gov/ohrp/regulations-and-policy/guidance/faq/2018-requirements-faqs/index.html

Office for Human Research Protections (OHRP) (n.d.) Vulnerable populations. Retrieved November 29, 2022, from https://www.hhs.gov/ohrp/regulations-and-policy/guidance/vulnerable-populations/index.html

Office International des Epizooties (OIE). (2018). Infection with Lumpy Skin Disease Virus. Terrestrial Animal Health Code – Article 11.9.1. Retrieved November 29, 2022, from https://www.oie.int/fileadmin/Home/eng/Health_standards/tahc/2018/en_chapitre_lsd.htm

Official Gazette. *Law on the Practice of Medicine and Medical Sciences*, 1219 § (1928). Retrieved May 22, 2022, from https://www.mevzuat.gov.tr/mevzuat?MevzuatNo=1219&MevzuatTur=1&MevzuatTertip=3

Official Gazette. *Basic Law on Health Care*, 3359 § (1987). Retrieved May 22, 2022, from https://www.mevzuat.gov.tr/mevzuat?MevzuatNo=3359&MevzuatTur=1&MevzuatTertip=5

Official Gazette. *Regulation on Patient Rights*, 23420 § (1998). Retrieved May 22, 2022, from https://www.mevzuat.gov.tr/mevzuat?MevzuatNo=4847&MevzuatTur=7&MevzuatTertip=5

Official Gazette. *Turkish Penalty Law*, 5237 § (2004). Retrieved May 22, 2022, from https://www.mevzuat.gov.tr/mevzuat?MevzuatNo=5237&MevzuatTur=1&MevzuatTertip=5

Official Gazette. *Biosafety Law*, 5977 § (2010). Retrieved May 22, 2022, from https://www.mevzuat.gov.tr/mevzuat?MevzuatNo=5977&MevzuatTur=1&MevzuatTertip=5

Oral History Association (2021). Oral history in the digital age. Retrieved November 27, 2022, from http://ohda.matrix.msu.edu/category/ethics/

Örnek Büken, N. (2018). Women in the principle of respect for human vulnerability and personal integrity. *Turkiye Klinikleri Journal of Obstetrics & Women's Health Disparities in Nursing—Special Topics* 4(1), 1–11.

Örnek Büken, N., & Küreci, H. D. (2018). Children as participants in biomedical research and conditions for researching in children in Turkey. *Turkiye Klinikleri Journal of Medical Ethics and Law History—Special Topics* 4(1), 50–60.

Overson Law, PLLC. (n.d.) What happens when you go to drug court in [state]? Retrieved July 4, 2022, from https://www.utahcriminallaw.net/happens-go-drug-court-utah/

Paasche-Orlow, M., Taylor, H., & Brancati, F. (2003). Readability standards for informed consent forms as compared with actual readability. *New England Journal of Medicine* 348(8), 721–726.

Paredes, P., de la Peña, M., Flores-Guerra, E., Diaz, J., & Trostle, J. (1996). Factors influencing physicians' prescribing behavior in the treatment of childhood diarrhea: Knowledge may not be the clue. *Social Science & Medicine*, 42(8), 1141–1153. https://doi.org/10.1016/02779536(95)00387-8

Patel, E.U., Solomon, S.S., Lucas, G.M., McFall, A.M., Srikrishnan, A.K., Kumar, M.S., Iqbal, S.H., Saravanan, S., Paneerselvam, N., Balakrishnan, P., Laeyendecker, O., Celentano, D.D., & Mehta, S.H. (2021). Temporal change in population-level prevalence of detectable HIV viraemia and its association with HIV incidence in key populations in India: A serial cross-sectional study. *Lancet HIV* 8: e544–53. https://doi: 10.1016/S2352-3018(21)00098-9

Patel, S., & Gadit, A.M. (2008). Karo-Kari: A form of honour killing in Pakistan. *Transcultural Psychiatry* 45(4), 683–694. doi:10.1177/1363461508100790

People's Republic of China State Food and Drug Administration (SFDA) (2003). Good clinical practice. Retrieved May 22, 2022, from http://www.nhc.gov.cn/yzygj/s7659/202004/ld5d7ea301f04adba4c4e47d2e92eb96.shtml.

Porcel, R.O. (2021). LSD epidemiological situation in Europe and neighbouring areas – Occurrence and vaccination. Current State of play (evolution since GF TADs – SGE LSD9). Retrieve November 29, 2022, from https://rr-europe.oie.int/wp-content/uploads/2021/01/lsd_epidemiological_situation_europe_dg-sante_sge-lsd10.pdf

Pratt, B., Sheehan, M., Barsdorf, N., & Hyder, A.A. (2018). Exploring the ethics of global health research priority-setting. *BMC Medical Ethics* 19(1), 94. doi:10.1186/s12910-018-0333-y

Presidency of the Republic of Turkey Regulatory Information System (RIS). (1928). Law on the Practice of Medicine and Medical Sciences, Law Number 1219. Retrieved May 22, 2022, from https://www.mevzuat.gov.tr/mevzuat?MevzuatNo=1219&MevzuatTur=1&MevzuatTertip=3

Presidency of the Republic of Turkey Regulatory Information System (RIS). (1987). Basic Law on Health Care, Law Number 3359. Retrieved May 22, 2022, from https://www.mevzuat.gov.tr/mevzuat?MevzuatNo=3359&MevzuatTur=1&MevzuatTertip=5

Presidency of the Republic of Turkey Regulatory Information System (RIS). (1998). Regulation on Patient Rights. Law Number 23420. Retrieved May 22, 2022, from https://www.mevzuat.gov.tr/mevzuat?MevzuatNo=4847&MevzuatTur=7&MevzuatTertip=5

Presidency of the Republic of Turkey Regulatory Information System (RIS). (2004). Turkish Penalty Law. Law Number 5237. Retrieved May 22, 2022, from https://www.mevzuat.gov.tr/mevzuat?MevzuatNo=5237&MevzuatTur=1&MevzuatTertip=5

Presidency of the Republic of Turkey: Regulatory Information System (RIS). (2010). Biosafety Law. Law Number 5977. Retrieved May 22, 2022, from https://www.mevzuat.gov.tr/mevzuat?MevzuatNo=5977&MevzuatTur=1&MevzuatTertip=5

Rankin, J. (2021). Hungary passes law banning LGBT content in schools or kids' TV. *Guardian*. Retrieved July 5, 2022, from https://www.theguardian.com/world/2021/jun/15/hungary-passes-law-banning-lbgt-content-in-schools

Ratzan, S.C., & Parker, R.M. (2000). Introduction: Health literacy implications of the Affordable Care Act. In National Library of Medicine Current Bibliographies in Medicine: Health Literacy. NLM Pub. No. CBM 2000-1. Retrieved May 22, 2022, from www.chcs.org/media/Health_Literacy_Implications_of_the_Affordable_Care_Act.pdf

Rawls, J. (1999). *A theory of justice*, revised edition. Boston: Harvard University Press.

Regitz-Zagrosek, V., & Seeland, U. (2012). Sex and gender differences in clinical medicine. *Handbook of Experimental Pharmacology* 214, 3-22. https://doi: 10.1007/978-3-642-30726-3_1

Renkin, H.Z. (2009). Homophobia and queer belonging in Hungary. *Focaal*, 2009(53), 20–37. https://doi.org/10.3167/fcl.2009.530102

Resnik, D.B. (2015). What is ethics in research & why is it important? Retrieved April 24, 2022, from https://www.niehs.nih.gov/research/resources/bioethics/whatis/index.cfm

Resnik, D.B., & Jones, C.W. (2006). Research subjects with limited English proficiency: Ethical and legal issues. *Accountability in Research* 12(2), 157–177. https://doi:10.1080/08989620600654043.

Roa, S. (2015). Responses to rape in schools. *Economic & Political Weekly* 50(3), online. Retrieved November 27, 2022, from https://www.epw.in/journal/2015/3/commentary/responses-rape-schools.html

Rodzinka, M. (2020). Mental health in all policies: Scoping review of national developments (2017-2019). Retrieved July 6, 2020, from https://www.mhe-sme.org/wp-content/uploads/2020/02/MHAP_Scoping_review_2017_2019_final.pdf

Roser, M., & Ortiz-Ospina, E. (2018). Literacy: Our world in data. Retrieved May 22, 2022, from https://ourworldindata.org/literacy

Rothstein, M. (2015). Ethical issues in Big Data Health Research. *Journal of Law, Medicine, & Ethics* 43(2), 425–429. doi:https://doi.org/10.1111/jlme.12258

Rothwell, E., Johnson, E., Wong, B., Goldenberg, A., Tarini, B.A., Riches, N., Stark, L.A., Pries, C., Langbo, C., Langen, E., & Botkin, J. (2020). Comparison of video, app and standard consent processes on decision making for biospecimen research: A randomized controlled trial. *Journal of Empirical Research on Human Research Ethics* 15(4), 252–260. doi:10.1177/1556264620913455

Roughgarden, J. (2013). *Evolution's rainbow: Diversity, gender, and sexuality in nature and people*. Berkeley, CA: University of California Press.

Sánchez, F.J., & Vilain, E. (2009). Collective self-esteem as a coping resource for male-to-female transsexuals. *Journal of Counseling Psychology*, 56, 202–209. https://doi.org/10.1037/a0014573

Sanmukhani, J., & Tripathi, C. B. (2011). Ethics in clinical research: The Indian perspective. *Indian Journal of Pharmaceutical Sciences* 73(2), 125–130. https://doi.org/10.4103/0250-474x.91564

Scientific American, The Editors. (2022). Clinical trials need more diversity. *Scientific American*, Special Editions 31, 3s, 50 (June 2022). https://doi:10.1038/scientificamericansocialjustice0622-50

ShareAction (2021). Reforming the rules. Retrieved July 4, 2022, from https://shareaction.org/what-we-do/reforming-the-rules

Sherin A. (2013). Research ethics committees (RECs) and monitoring of biomedical research in Pakistan. *Khyber Medical University Journal 5*(3), 121–122.

Siegel, L. (2018). Health literacy and the IRB: Considering literacy while conducting medical and public health research: Informed consent procedures. Retrieved May 22, 2022, from https://www.hsph.harvard.edu/healthliteracy/health-literacy-and-the-irb/

Smeaton, L.M., Kacanek, D., Mykhalchenko, K., Coughlin, K., Klingman, K.L., Koletar, S.L., Barr, E., & Collier, A.C. (2020). Screening and enrollment by sex in Human Immunodeficiency Virus Clinical Trials in the United States. *Clinical Infectious Diseases,* 71(5), 1300–1305. doi:10.1093/cid/ciz959

Solar, O., & Irwin A. (2010). A conceptual framework for action on the social determinants of health. Retrieved November 27, 2022, from https://www.who.int/publications/i/item/9789241500852

Srivastava, A. (2020). Guidelines for ethical considerations in social research & evaluation in India. Retrieved November 28, 2022, from https://cmsindia.org/sites/myfiles/Guidelines-for-Ethical-Considerations-in-Social-Research-Evaluation-In-India_2020.pdf

Srivastava, K., Chatterjee, K., & Bhat, P.S. (2016). Mental health awareness: The Indian scenario. *Industrial Psychiatry Journal* 25(2), 131–134. https://doi.org/10.4103/ipj.ipj_45_17

Steinberg, J.R., Turner, B.E., Weeks, B.T., Magnani, C.J., Wong, B.O., Rodriguez, F., Yee, L.M., & Cullen, M.R. (2021). Analysis of female enrollment and participant sex by burden of disease in US clinical trials between 2000 and 2020. *JAMA Network Open*, 4(6), e2113749. doi:10.1001.jamanetworkopen.2021.13749

Stubnya, G., Nagy, Z., Lammers, C.H., Rihmer, Z., & Bitter, I. (2010). Deinstitutionalization in Europe: two recent examples from Germany and Hungary. *Psychiatria Danubina* 22(3), 406–412.

Svensson, S., Balogh, P., & Cartwright, A. (2017). Solidarity in the treatment of mental illness in Hungary: A case study of the Awakenings Foundation as a vehicle for change. CEU Center for Policy Studies. Retrieved December 27, 2021, from https://cps.ceu.edu/sites/cps.ceu.edu/files/attachment/publication/2947/cps-working-paper-solidus-solidarity-mental-illness-treatment-2017_2.pdf

Szel, Z., Kiss, D., Torok, Z., Gyarmathy, A. (2020). Hungarian medical students' knowledge about and attitude toward homosexual, bisexual, and transsexual individuals. *Journal of Homosexuality* 67(10), 1429–1446. https://doi.org/10.1080/00918369.2019.1600898

Taş, M., Balcı, M., Yüksel, A., & Şahin Yeşilbuçuk, N. (2015). Consumer awareness, perception and attitudes towards genetically modified foods in Turkey. *British Food Journal* 117(5), 1426–1439. https://doi.org/10.1108/BFJ-01-2014-0047

Tavake-Pasi, O.F., Napia, E., Simonsen, S., Davis, F., Ralls, B., Mukundente, V., Coulter, H., Sanchez-Birkhead, A., Eisenman, P., Alder, S., Toki, U., Stark, L., Digre, K., Aiono, E., Johnston, L., & Sunada, G. (2016). Native Hawaiian Pacific Islanders: The ripple effect improving obesity-related behaviors through partnerships and community wellness coaches. Poster presentation at the American Public Health Association Meetings in Denver, Colorado.

Tealdi, J.C. (2015). Problemas bioéticos en la investigación de nuevas vacunas: ¿obedecen a razones de salud pública? *Salud colectiva* 11(1), 87–97. Retrieved July 11 2022, from http://www.scielo.org.ar/scielo.php?script=sci_arttext&pid=S1851-82652015000100007&lng=es&tlng=es.

Thatte, U.M. (2002). Do all projects require ethics committee clearance? *Journal of Postgraduate Medicine* 48(2), 91. https://www.jpgmonline.com/text.asp?2002/48/2/91/139

Thorat, S., & Verma, S. (2017). Social science research in India: Status, issues, and policies. *Oxford Scholarship Online*. https://doi.org/10.1093/acprof:oso/9780199474417.001.0001

Times of India, The. (2020). International Literacy Day 2020: How literacy rate is calculated in India: India News. *The Times of India*. Retrieved December 6, 2021, from https://timesofindia.indiatimes.com/india/international-literacy-day-2020-how-literacy-rate-is-calculated-in-india/articleshow/77991200.cms

Transvanilla. (2020). The Hungarian government moves to ban legal gender recognition. Retrieved April 1, 2020, from https://transvanilla.hu/home/news/hungarian-government-bans-legal-gender-recognition

TRUST. (2018). Global Code of Conduct for Research in Resource-Poor Settings. doi:10.48508/GCC/2018.05. Retrieved November 30, 2022, from https://www.globalcodeofconduct.org/

Turkish Ministry of Health: Drug and Medical Device Agency (2013). Regulation on clinical trials of drugs and biological products. Law Number 28617. Retrieved May 22, 2022, from https://www.mevzuat.gov.tr/mevzuat?MevzuatNo=17285&MevzuatTur=7&MevzuatTertip=5

Turkish Ministry of Health: Drug and Medical Device Agency (TİTCK). (2015). Guideline on ethical approaches for the clinical studies on pediatric populations. Retrieved May 22, 2022, from https://www.titck.gov.tr/mevzuat/2382

Ugalde, A. & Homedes, N. (2013). La contribución de las revistas clínicas de alto impacto a la ciencia: el caso de The Lancet. *Salud Colectiva* 9(1), 5–10.

UNESCO (2013). The principle of respect for human vulnerability and personal integrity: Report of the International Bioethics Committee of UNESCO (IBC). Paris: UNESCO. Retrieved May 22, 2022, from https://unesdoc.unesco.org/ark:/48223/pf0000219494

United Nations. (1995). Convention on the Rights of the Child. UNHCR. Retrieved May 22, 2022, from https://www.ohchr.org/en/professionalinterest/pages/crc.aspx.

United Nations (2018a). Convention on the Rights of the Child. UNHCR. Retrieved May 22, 2022, from https://en.wikipedia.org/wiki/Convention_on_the_Rights_of_the_Child

United Nations (2018b). Universal Declaration of Human Rights (UDHR). Retrieved April 24, 2022, from http://www.un.org/en/universal-declaration-human-rights/

United Nations. (n.d.). Sustainable Development Goals. Retrieved August 3, 2022, from https://sdgs.un.org/goals

United States Congress (2019). Newborn screening saves lives Reauthorization Act of 2019. Retrieved June 29, 2022, from https://www.congress.gov/congressional-report/116th-congress/house-report/174/1

United States Department of Health, Education, and Welfare (1979). The Belmont Report. Retrieved June 22, 2022, from https://www.hhs.gov/ohrp/sites/default/files/the-belmont-report-508c_FINAL.pdf

United States Department of Health and Human Services (USDHHS) (2017). The Belmont Report. Retrieved April 24, 2022, from https://www.hhs.gov/ohrp/regulations-and-policy/belmont-report/index.html

United States Department of Health and Human Services (USDHHS) (2019). Definitions of research and human subjects. Retrieved April 24, 2022, from https://www.hhs.gov/programs/research/index.html.

United States Department of Health and Human Services (USDHHS) (2020). Code of federal regulations (annual edition). Title 45: Public Welfare. Subpart A: Basic HHS Policy for Protection of Human Research Subjects. Retrieved May 22, 2022, from https://www.govinfo.gov/content/pkg/CFR-2016-title45-vol1/pdf/CFR-2016-title45-vol1-part46.pdf.

United States Department of Health and Human Services (USDHHS) (2022). HIV treatment as prevention. Retrieved August 11, 2022, from https://www.hiv.gov/tasp

United States Department of Veterans Affairs (USDVA) (2016). Epidemiology Research Studies (ERS). Retrieved April 24, 2022, from https://www.publichealth.va.gov/epidemiology/studies/index.asp

United States Holocaust Memorial Museum (USHMM) (2019a). Nazi Medical Experiments. Retrieved April 24, 2022, from https://www.ushmm.org/wlc/en/article.php?ModuleId=10005168

United States Holocaust Memorial Museum (USHMM) (2019b). Nuremberg Code. Retrieved April 24, 2022, from https://www.ushmm.org/information/exhibitions/online-exhibitions/special-focus/doctors-trial/nuremberg-code

United States National Institutes of Health (1994). NIH guidelines on the inclusion of women and minorities as subjects in clinical research. Retrieved November 27, 2022, from https://grants.nih.gov/grants/guide/notice-files/not94-100.html

United States Office for Human Research Protections (OHRP). (n.d.) Prisoner research FAQs. Retrieved July 4, 2022, from https://www.hhs.gov/ohrp/regulations-and-policy/guidance/faq/prisoner-research/index.html

United States Office for Human Research Protections (OHRP). (2008). Coded private information or specimens use in research—guidance. Retrieved November 29, 2022, from https://www.hhs.gov/ohrp/regulations-and-policy/guidance/research-involving-coded-private-information

United States Office for Human Research Protections (OHRP). (2016). 45 CFR 46. 16 Feb 2016. Retrieved July 4, 2022, from https://www.gpo.gov/fdsys/pkg/CFR-2016-title45-vol1/pdf/CFR-2016-title45-vol1-part46.pdf

United States Office of Human Research Protections (OHRP). (2018). The Belmont Report. Retrieved May 22, 2022, from https://www.hhs.gov/ohrp/regulations-and-policy/belmont-report/read-the-belmont-report/index.html#xjust

Untas, A., Thumma, J., Rascle, N., Rayner, H., Mapes, D., Lopes, A.A., Fukuhara, S., Akizawa, T., Morgenstern, H., Robinson, B.M., Pisoni, R.L., & Combe, C. (2011). The associations of social support and other psychosocial factors with mortality and quality of life in the dialysis outcomes and practice patterns study. *Clinical Journal of the American Society of Nephrology: CJASN* 6(1), 142–152. https://doi.org/10.2215/CJN.02340310

van Diemen, J., Verdonk, P., Chieffo, A., Regar, E., Mauri, F., Kunadian, V., Sharma, G., Mehran, R., & Appelman, Y. (2021). The importance of achieving sex- and gender-based equity in clinical trials: A call to action. *European Heart Journal* 42, 2990–2994. doi:10.1093/eurheartj/ehab457

Van Maren, J. (2018). Doctors propose killing sick children under Canada's new euthanasia law. Life Site-Bioethics, End of Life. Retrieved November 27, 2022, from https://www.lifesitenews.com/blogs/doctors-propose-killing-sick-children-under-canadas-new-euthanasia-law

Vijayananthan, A., & Nawawi, O. (2008). The importance of Good Clinical Practice guidelines and its role in clinical trials. *Biomedical Imaging and Intervention Journal* 4(1), e5. doi: 10.2349/biij.4.1.e5. https://www.ncbi.nlm.nih.gov/pmc/articles/PMC3097692/

Wade, L. (2018). To overcome decades of mistrust, a workshop aims to train Indigenous researchers to be their own genome experts. Retrieved November 27, 2022, from Science. doi:10.1126/science.aav5286.

West-Oram, P.G.N., Brooks, C., Jenkins, V. (2020). From subject to fellow researcher: Reconceptualising research relationships to safeguard potentially vulnerable survey participants. *The American Journal of Bioethics* 20(10), 72–74. https://doi.org/10.1080/15265161.2020.1806385

Williams, J.R. (2008). The Declaration of Helsinki and Public Health. *Bulletin of the World Health Organization* 86(8), 650–652. doi:10.247/BLT.08.050955.

World Bank. (2012). HIV/AIDS in India. Retrieved August 9, 2022, from https://www.worldbank.org/en/news/feature/2012/07/10/hiv-aids-india#

World Bank Country Reports (n.d.). Morocco. Retrieved November 27, 2022, from https://data.worldbank.org/country/morocco

World Economic Forum. (2021). Global gender gap report 2021. Retrieved from https://www.weforum.org/reports/global-gender-gap-report-2021/in-full/gggr2-benchmarking-gender-gaps-findings-from-the-global-gender-gap-index-2021

World Health Organization. (2007). Everybody's business: Strengthening health systems to improve health outcomes: WHO's framework for action. Retrieved June 22, 2022, from https://apps.who.int/iris/handle/10665/43918

World Health Organization (WHO) (2017). Diarrhoeal disease. Retrieved May 22, 2022, from https://www.who.int/news-room/fact-sheets/detail/diarrhoeal-disease.

World Health Organization. (2021). Micronutrients. Retrieved November 27, 2022, from https://apps.who.int/nutrition/topics/micronutrients/en/index.html

World Medical Association (2008). Declaration of Helsinki – Ethical Principles for Medical Research Involving Human Subjects. Retrieved November 27, 2022, from https://www.wma.net/policies-post/wma-declaration-of-helsinki-ethical-principles-for-medical-research-involving-human-subjects/

World Medical Association. (various dates, 1964–2008). Declaration of Helsinki. Retrieved June 29, 2022, from https://www.wma.net/what-we-do/medical-ethics/declaration-of-helsinki/

World Organisation for Animal Health (2021). GF-TADs-Standing Group of Experts on Lumpy Skin Disease in South-east Europe. Retrieved November 29, 2022, from https://rr-europe.oie.int/en/Projects/gf-tads-europe/2-standing-groups-of-experts-on-lumpy-skin-disease-in-south-east-europe/

Yount, K.M. (2003). Provider bias in the treatment of diarrhea among boys and girls attending public facilities in Minia, Egypt. *Social Science & Medicine* 56(4), 753–768. https://doi.org/10.1016/s0277-9536(02)00072-2

Zavaleta, N., Figueroa, D., Rivera, J., Sanchez, J., Alfaro, S., & Lonnerdal, B. (2007). Efficacy of rice-based oral rehydration solution containing recombinant human lactoferrin and lysozyme in Peruvian children with acute diarrhea. *Journal of Pediatric Gastroenterology & Nutrition* 44 (2), 258–264. https://doi.org/10.1097/mpg.0b013e31802c41b7.

Zubair, S., & Zubair, M. (2017). Situating Islamic feminism(s): Lived religion, negotiation of identity and assertion of third space by Muslim women in Pakistan. *Women's Studies International Forum* 63, 17–26. doi:10.1016/j.wsif.2017.06.002

Žvan-Elliott, K. (2015). *Modernizing patriarchy: The politics of women's rights in Morocco*. Austin, TX: University of Texas Press.

Index

Global Political Justice

Scholarly debate on the subject of global justice has been overwhelmingly focused so far on the *socio-economic* aspects of justice. Much less attention has been given to those *political* aspects of global justice concerned with arrangements for public decision-making and the collective exercise and control of power. This gap is not adequately filled by literatures on global democracy, either, since these do not incorporate sufficient analysis of whether the democratic institutions that deliver political justice within states can achieve the same result when dealing with the very different forms of power and political agency that structure the domain of global politics.

This collection brings together scholars from across the disciplines of political theory, normative ethics, and International Relations to undertake a fresh examination of some fundamental theoretical questions about the nature and significance of global political justice. Contributors tackle several dimensions of this complex theoretical topic, exploring questions about: the relationship of global political justice to other normative standards like 'legitimacy', 'democracy', and 'socio-economic' justice; the nature of global 'public power' and the prospects for global political community; the justice and continued significance of traditional ordering principles of sovereignty and territoriality; and the relevance of standards of political justice (like political equality) to allocations of burdensome political responsibilities, such as those associated with ideas of 'climate debt'.

This book was originally published as a special issue of *Critical Review of International Social and Political Philosophy.*

Terry Macdonald is a Senior Lecturer in International Relations at the University of Melbourne, Australia.

Miriam Ronzoni is the Principal Investigator of the Sofja Kovalevskaja Research Project "Background Justice between States" at Technical University of Darmstadt.